OXFORD MEDICAL PUBLICATIONS

Eczema in Childhood

THE FACTS

ALSO PUBLISHED BY OXFORD UNIVERSITY PRESS

Ageing: the facts (second edition)
Nicholas Coni, William Davison, and Stephen Webster

Alcoholism: the facts (second edition)
Donald W. Goodwin

Allergy: the facts
Robert Davies and Susan Ollier

Arthritis and rheumatism: the facts (second edition)
J. T. Scott

Autism: the facts
Simon Baron-Cohen and Patrick Bolton

Back pain: the facts (third edition)
Malcolm I. V. Jayson

Bowel cancer: the facts
John M. A. Northover and Joel D. Kettner

Contraception: the facts (second edition)
Peter Bromwich and Tony Parsons

Coronary heart disease: the facts (second edition)
Desmond Julian and Claire Marley

Cystic fibrosis: the facts (second edition)
Ann Harris and Maurice Super

Deafness: the facts
Andrew P. Freeland

Dyslexia and other learning difficulties: the facts
Mark Selikowitz

Eating disorders: the facts (third edition)
Suzanne Abraham and Derek Llewellyn-Jones

Head injury: the facts
Dorothy Gronwall, Philip Wrightson, and Peter Waddell

Healthy skin: the facts
Rona M. MacKie

Kidney disease: the facts (second edition)
Stewart Cameron

Liver disease and gallstones: the facts (second edition)
Alan G. Johnson and David R. Triger

Lung cancer: the facts (second edition)
Chris Williams

Multiple sclerosis: the facts (third edition)
Bryan Matthews

Obsessive–compulsive disorder: the facts
Padmal de Silva and Stanley Rachman

Parkinson's disease: the facts (second edition)
Gerald Stern and Andrew Lees

Pre-eclampsia: the facts
Chris Redman and Isabel Walker

Thyroid disease: the facts (second edition)
R. I. S. Bayliss and W. M. G. Tunbridge

Eczema in Childhood

THE FACTS

DAVID J. ATHERTON

Consultant in Paediatric Dermatology
Hospital for Sick Children
Great Ormond Street, London

OXFORD NEW YORK TOKYO
OXFORD UNIVERSITY PRESS
1994

Oxford University Press, Walton Street, Oxford OX2 6DP

Oxford New York Toronto
Delhi Bombay Calcutta Madras Karachi
Kuala Lumpur Singapore Hong Kong Tokyo
Nairobi Dar es Salaam Cape Town
Melbourne Auckland Madrid

and associated companies in
Berlin Ibadan

Oxford is a trade mark of Oxford University Press

Published in the United States
by Oxford University Press Inc., New York

A catalogue record for this book is available from the British Library

Library of Congress Cataloging in Publication Data

ISBN 0 19 261351 0 (Hbk)
ISBN 0 19 262398 2 (Pbk)

Typeset by Footnote Graphics, Warminster, Wilts
Printed by Interprint, Malta

Preface

Childhood eczema is unfortunately very common. Although it is generally a relatively minor problem, some children endure prolonged and almost unbearable suffering. The pain caused by this disease to sufferers, and to their close relatives, is little appreciated, and the level of help available to both leaves much to be desired. In the course of my work as a paediatric dermatologist, I see many hundreds of eczematous children and their parents each year. This book has been written to provide information about the disease to parents who have a child with eczema to care for, in the hope that it will make their lives a little easier.

Parents often complain to me about their experiences with doctors, and no doubt they sometimes complain to others about me in the same way. Doctors are at their happiest dealing with diseases that respond predictably and completely to treatment, and this is understandable. Unfortunately, severe childhood eczema is not one of these diseases. Parents often feel that the child's doctors have abandoned them to the disease. They often say that they have been told that 'nothing can be done' for eczema. Nothing could be further from the truth. There is always something that can be done, and all good doctors know this. Looking after children with severe eczema can be very difficult, demanding and frustrating for a doctor, but it can also be very rewarding, for the same reasons. By understanding the disease a little better, I hope the reader will also see the doctor's problem more clearly, what he can do and what he cannot. I also hope that this will help parents and doctors to work together as a team in caring for children with eczema.

No doctor, however much he would like, has the time to cover with you more than a fraction of what is in this book. It contains a mixture of facts, half-facts, and hypotheses, and I have tried to distinguish these as far as possible. Inevitably, what I have written reflects a rather personal view of atopic eczema, and other experts will certainly not agree with me on every point. This is of course unavoidable, and I make no apology for it, but it is something you should bear in mind both while reading it and when you subsequently see your child's own doctor. He may not see eczema exactly as I do, and he may be right.

With childhood eczema, the keynote is optimism: optimism because this condition, dreadful as it may be, almost always fades away in time, leaving little or no trace either physical or psychological, and optimism because research constantly brings us closer to understanding the nature of the disease, with the promise that, in the not too distant future, it may be prevented altogether.

I sincerely hope you find something worthwhile in this book. If you do, not one of those Sunday mornings spent at my desk will have been wasted.

London D.J.A.
April 1994

Acknowledgements

I am grateful to my wife Anne for her immense patience during the writing of this book. I am also grateful to Mrs Lesley Haynes for advice on elimination diets, to Mr David Taylor for comments on the section on eye problems, and to Mr P. Murtagh for his comments on Chapter 12. I thank the photographers at the Hospital for Sick Children, Great Ormond Street and at St John's Dermatology centre, St Thomas's Hospital for the excellent photographs of children in my care, and their parents for allowing me to use pictures of their children. Merck Dermatology provided financial support for the colour illustrations, which increases their value greatly and which I know readers will appreciate as much as I do.

Contents

1

The normal skin

Among the organs of the human body, the heart and the brain are the superstars, while the poor old skin is always cast in a supporting role. Yet in its own way, the skin is just as important. For a start, it is by far the largest organ in the body. Like the heart and the brain, it is essential for life. If just a third of the skin surface of a child is burnt, survival is unlikely despite the best medical treatment. The difference between the superstar organs and the skin is that the skin is generally more reliable, and only very rarely do skin diseases so upset the function of the skin that our lives are at risk.

The principal role of the skin is of course to provide us with a flexible protective covering. Perhaps the most vital of all the skin's protective functions is the conservation of water; it prevents us from literally drying out. It also provides a barrier that prevents our bodies from being invaded by the many bacteria, fungi, and other micro-organisms that lie in wait in our environment. The other important function of the skin is its role in the regulation of our body temperature. Humans, like other mammals, are designed to operate at a fixed temperature, and the skin plays the leading role in making short-term adjustments to keep the internal temperature constant.

Let's take a closer look at how the skin is designed to perform these functions. *Figure 1* illustrates the construction (the *anatomy*) of the skin. It comprises three main layers: the *epidermis*, the *dermis*, and the *subcutaneous fat*.

The epidermis is the outermost layer, and is responsible for most of its protective functions. There is a bottom (*basal*) layer of cells that are responsible for the production of new cells, their offspring continuously moving upwards toward the surface. On their way to the surface, these cells mature. As they mature, they become tough, flat, and waxy. Their toughness largely results from the internal production of large amounts of a protein called *keratin*. When they reach the surface, they take the form of flattened, overlapping plates, packed together tightly. These plates are bonded together by an oily, water-repellent cement, which is secreted from the cells as they mature. The plate-like cells are continually removed from the surface by being rubbed off, as fast as they are being replaced from below. The technical name for this outermost

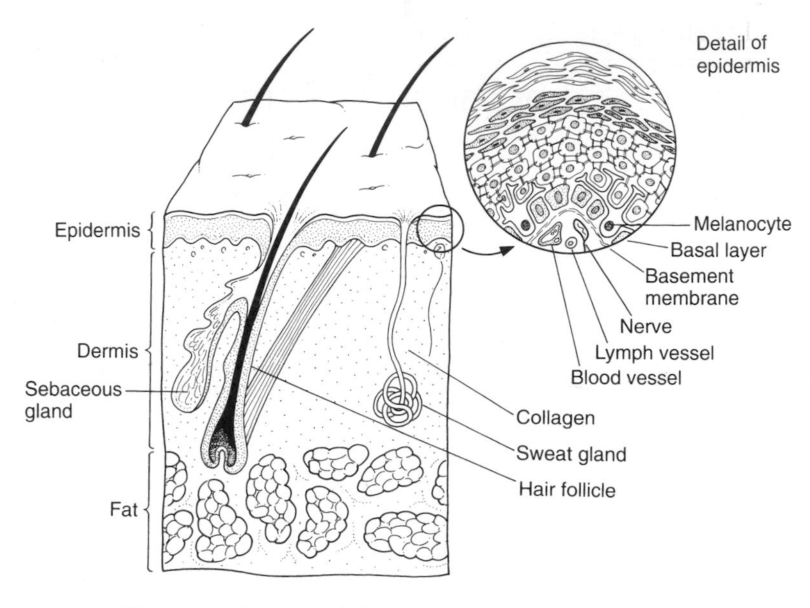

Figure 1 A diagram of skin as seen under the microscope.

layer is the *stratum corneum*. It is remarkably flexible, more or less waterproof, and has a dry surface that is inhospitable to micro-organisms. The surface is punctured by small channels through which hairs and sweat pass. An oily secretion called *sebum* also reaches the surface through the hair channels. Sebum is not secreted until puberty, and its functions remain somewhat unclear. It lubricates the hairs as they slide up through the skin, and adds an extra waterproofing layer to the skin surface, but probably its benefit derives principally from its antibacterial and antifungal effects.

The cooling type of sweat is called *eccrine* sweat, and this is produced on the palms and soles under emotional stimuli, and elsewhere mainly by thermal stimuli, i.e. when the body is becoming too hot. The main purpose of sweat is of course to provide a cooling effect, but on the palms and soles it has the important function of improving grip.

There is a second, less important type of sweat called *apocrine* sweat. This, like sebum, is not produced until puberty, and is made in glands which are restricted to certain sites, particularly the axillae and genital areas. Apocrine sweat is a milky fluid which decomposes to give the characteristic unpleasant odour with which most readers will be familiar.

The function of this secretion is unknown, but the main stimulus for its production is known to be sexual arousal, and it is widely believed that in its fresh state it contains aromas which have a seductive effect on members of the opposite sex who get close enough. Such aromas are known scientifically as *pheromones*. Interestingly, the glands in the mammalian breast that secrete milk are in fact giant and somewhat modified apocrine glands. This is why the smell of ripe Camembert and overripe armpits are so similar.

Beneath the epidermis is the *dermis*. The dermis exists solely to support the epidermis. It consists largely of a meshwork of strong fibres, made from a protein called *collagen*, which provide most of the tensile strength of the skin. Through this meshwork run the blood vessels that nourish the epidermis, and the nerves that provide it with the ability to sense temperature changes and pain. These nerves are of great importance in keeping the brain informed of conditions outside the body and on the surface of the skin. The combination of waterproofing, strength, flexibility, and extreme sensitivity is the great design achievement of the skin.

As well as blood vessels and nerves, the dermis contains a network of tiny channels through which passes a fluid called *lymph*. Lymph is the clear fluid that bathes all our tissues; we are perhaps most aware of it when it leaks on to the surface following a graze. It is formed by a process of continuous seepage out of small blood vessels, into the surrounding tissues. From here it is collected into tiny lymph channels (*lymphatics*), in which it travels into progressively larger and larger channels until it is returned into the bloodstream. On its way, it is filtered to remove microscopic debris and micro-organisms it may have picked up. These filters are small structures called *lymph nodes* or *lymph glands*, which one can normally feel as small lumps at certain sites, such as under the arms and in the groins.

Apart from providing a protective barrier, perhaps the skin's most important function is its role in temperature regulation. In some respects, the skin resembles the radiator of a car; it is responsible for dissipating excess heat in order to prevent a rise in temperature. The body's heating system itself has much in common with an ordinary domestic central heating system. The body's boiler is the liver, and the food we eat is the fuel. The boiler setting can be varied. The pipes are the blood vessels, and the radiators are the skin. To vary the heat output of the skin, the flow of blood through the skin can be varied. When the skin blood flow is increased, the skin becomes red. When it is decreased, the skin becomes white or even blue.

Beneath the dermis is the *subcutaneous fat*. This fat acts as an insulating layer which keeps the core of the body warm when heat is to be conserved. It also acts as a store of fuel for emergency use.

So the skin is not just 'hide'; it plays an important, active role in the team of organs even if it is not a glamorous one; it is the right back rather than the centre forward. Although skin diseases rarely interfere with its functions to the point at which life cannot be sustained, lesser degrees of interference are not uncommon. Eczema will, for example allow increased water loss through the skin by evaporation, and parents often notice that their child drinks a great deal to compensate. They sometimes worry that their child might have something wrong (like diabetes) to cause this thirst, but unless the child is producing more urine than is normal, there is no cause for anxiety.

Temperature control is another function of the skin which is often upset in children with eczema. They seem to fluctuate wildly between feeling too hot and too cold. The problem stems from the fact that the inflammation of the eczema causes an increase in skin blood flow, and hence the loss of excessive amounts of heat. This increased heat loss would lead to a fall in body temperature if there were no compensatory mechanisms. However, in practice the body responds automatically to the increased heat loss by turning up the 'boiler'; it increases heat output from the liver. This requires an increased supply of fuel, which, in its turn, means a greater intake of food, or using up body fat stores. A hearty appetite is common in children with eczema, and few are overweight.

The increased heat output in children with eczema can itself cause problems, however. If a child whose own internal heat production is increased goes into a hot environment, there may be difficulty dissipating the extra heat, as heat loss through the skin is already at a very high level. This difficulty in getting rid of extra heat is aggravated by the fact that sweating is impaired in skin affected by eczema, due to damage to the sweat channels. Most children with more severe degrees of eczema do have difficulty sweating, as well as a greatly increased internal heat production, so that they tend to 'overheat'. Such children generally therefore prefer a cooler environment than most normal individuals, and like less bedclothes at night. These are the reasons for the sudden fluctuations eczematous children may experience between feeling too hot and too cold. These fluctuations do not have serious consequences, but do cause very considerable discomfort.

2

What is eczema?

The skin can be injured directly by external factors, or it may be injured by events occurring within the body, though these may themselves be triggered by outside influences.

'Eczema' is a label which doctors apply to a particular pattern of reaction of the skin to injury. A useful analogy would be a boiling car radiator. This is only one of many problems one can have with a car. Similarly, eczema is only one of many problems one can have with one's skin. A boiling radiator is easily recognized—the problem is to know why it is boiling. There are many reasons why it may have done so; similarly there are many causes of eczema. Because *people* make cars, we understand the way they work, and we can usually pinpoint the problem when something is wrong. Unfortunately, the construction of the human body is much more complex, and, since it wasn't designed by human beings (thank goodness!), we are at a distinct disadvantage when it comes to understanding exactly how it works and why it sometimes goes wrong. So here the analogy ends.

As it happens, the word 'eczema' actually comes from the Greek word for 'boiling', and the eczema reaction in the skin is recognized by the effects of excessive amounts of fluid initially accumulating in the epidermis (see p. 1) and then, when it is severe, forcing its way through to the surface (*Figure* 2). I will consider this reaction in more detail later on (see p. 24).

The kind of eczema we perhaps understand best is called 'contact' eczema, or contact 'dermatitis'—note that the words eczema and dermatitis are used interchangeably. We know that many substances will cause an eczematous reaction if they are applied to the skin of otherwise normal people. Such substances include very powerful acids, solvents, and detergents. Anyone who regularly gets enzyme-containing washing powder on to wet hands will get eczema on them sooner or later. Another example is napkin dermatitis (nappy rash), which appears to be a reaction of very young, sensitive skin to urine and faeces. Individual susceptibility to these types of reaction—known as 'irritant' contact dermatitis—does of course vary quite a lot. One baby may get nasty nappy rash, and the next, though treated in exactly the

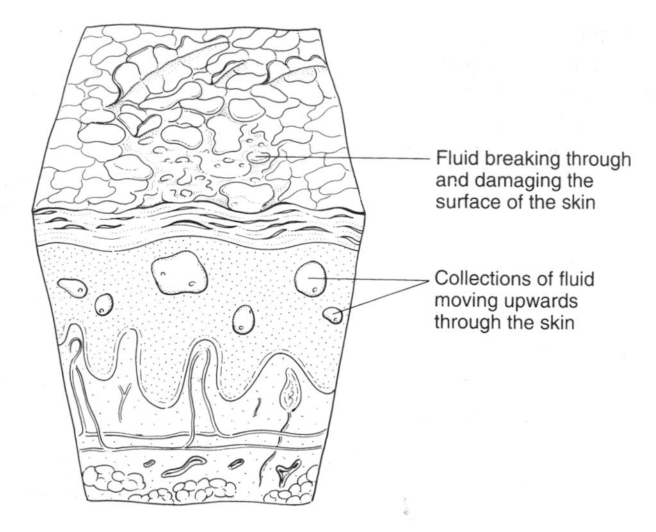

Figure 2 The appearance of eczematous skin under the microscope.

same way, may get no sign of it whatsoever. A second type of contact dermatitis is caused not by an irritant effect, but by an allergic reaction of the skin to an otherwise harmless substance applied to it; this type is called 'allergic contact dermatitis'. Nickel is a good example of the kind of substance that can cause allergic contact dermatitis. This metal is used to make cheaper jewellery, fasteners on clothing, and is also used in coins. Some people become allergic to it, and from then on will develop eczema whenever and wherever they come into contact with it. The mechanism by which such allergies arise is well established, and we have reliable tests, called 'patch' tests, to identify them.

Unfortunately, we understand much less clearly the causes of other types of eczema. Doctors give different names to the various forms of eczema they can recognize; these include 'seborrhoeic' dermatitis, 'nummular' eczema, and 'atopic' eczema. These various forms of eczema are distinguished mainly by the parts of the body they affect; it is important to attempt this distinction because it allows the doctor to know whether it is likely that the cause can be identified. For example, a good dermatologist will recognize that a gardener with eczema on his face and a patch of eczema on one thigh is likely to have developed allergic contact dermatitis to a chemical in the red tips of a particular brand of matches. The matchbox is kept in his trouser pocket, and the vapour from the tip envelops his face whenever the matches are struck.

The eczema can be cured simply by changing to a different brand of matches. If a middle-aged man comes to the clinic with eczema consisting of small disc-like patches on the outside of the arms and legs, the dermatologist will recognize the pattern as that of 'nummular' eczema (from the Latin word *nummus*: a coin). He knows from this pattern that the man's eczema has not been caused by contact with irritants or by allergy to externally applied substances. He also knows that this type of eczema is difficult to eradicate permanently by any kind of treatment. The cause of nummular eczema is almost entirely unknown, but it does generally tend to disappear after a few years. When a four-year-old child comes in with eczema in the crook of the elbows, on the wrists, the backs of the knees, and around the ankles, he knows that the child has the type of eczema known throughout the world as 'atopic' eczema or 'atopic' dermatitis. This disease forms the subject of this book.

3

Why atopic?

The term *atopic* (meaning *alien* in Greek) was coined over 60 years ago to describe collectively a group of diseases that include asthma, hay-fever, and a particular type of eczema. It was appreciated that these conditions were not inherited in any very predictable or regular way, but it was also clear that they did tend to occur, either separately or together, in families whose individuals showed a definitely increased susceptibility. It was recognized that what was inherited was a *tendency* to become allergic to common and generally harmless environmental substances such as pollens. This tendency was, and still is, called *atopy*. We now know that this tendency to become allergic occurs because the atopic individual too readily produces antibodies of a type called immunoglobulin E, known for short as *IgE* (pronounced *eye—gee—ee*).

We can recognize that individuals are atopic by using tests that detect these IgE antibodies. There are several tests that are able to do this; in practice, the simplest is the so-called *prick test* (*Figure 3*).

In the prick test, a drop of a solution containing a common environmental substance (such as grass or tree pollens, house dust mites, moulds, cat and dog dander, egg, or milk) is placed on the skin. A needle is then pricked very carefully into the outermost part of the skin through the drop, thereby introducing a minute amount of the test solution into the skin.

Antibodies react very specifically with the substance against which they have been made, and not with anything else. If the skin contains IgE antibody against the substance that has been pricked into it, it reacts by producing a weal. The diameter of this weal can easily be measured, and this measurement is an index of the amount of IgE antibody that has been made against a particular substance. The substances used for prick tests all contain proteins, and it is these proteins that react with the antibodies. Proteins which react with antibodies are called *antigens*. The antigens used for skin testing are ones that most of us encounter many, many times in our daily lives.

About one in every three people will have a positive test to at least one everyday substance. However, although all these people are therefore atopics, many of them will *never* develop eczema, asthma, or

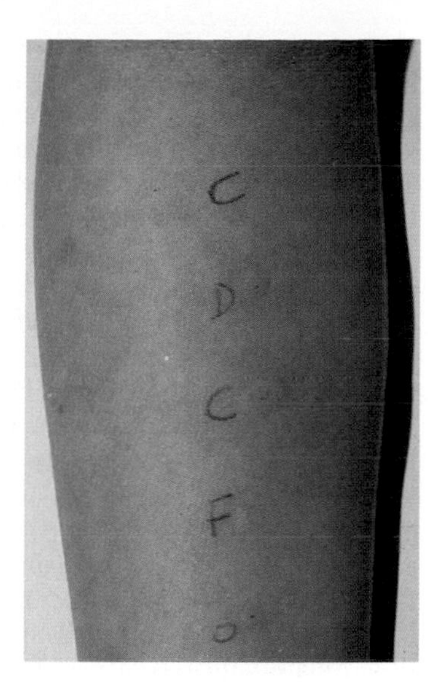

Figure 3 Prick tests.

hayfever. On the other hand, virtually everyone who *does* develop one of these diseases will be among this third of the population.

Atopic eczema is the type of eczema that occurs in atopics. Just as the name 'Ford' allows us to identify a particular type of car, so the adjective *atopic* allows us to identify a particular type of eczema, as well as telling us something about the person who has the eczema.

Our bodies need ways to defend themselves against their many natural enemies. Among the worst of these is a vast variety of viruses, bacteria, moulds, and parasites, many of which would kill us if given half a chance. We have several types of defence against this bunch of nasties. As well as antibodies, the blood contains special cells called *lymphocytes*, others called *neutrophils*, and yet others known as *macrophages*. These cells are known collectively as *white cells* (to distinguish them from the *red cells* in the blood, whose function is to transport oxygen). The antibodies and white cells are our soldiers, tanks, ships, and planes. Just as in wartime in the real world, the use of all these different elements of defence requires careful coordination. Each element has a special part to play, in cooperation with the others, and the loss of any one would result in increased vulnerability.

So antibodies are just one of the many forms of defence we have against intruders. They are themselves manufactured by cells known as *plasma cells*, and they react against antigens, which are almost always proteins. Viruses, bacteria, fungi, and parasites are made from proteins which are slightly different from the proteins from which we ourselves are made. The body is able to recognize these 'foreign' proteins, just as there are ways of recognizing other countries' soldiers, tanks, planes, and missiles. Once a protein is recognized as foreign—that is, as *antigenic*—the plasma cells will make antibodies against it. Antibodies are also called *immunoglobulins*—their chemical name. There are several different types of antibodies, known by code names: 'A', 'E', 'G', 'M', and so on. These different types of antibody have different ways of reacting with foreign proteins. Some form a coating on the protein, making it attractive to the cells known as macrophages, which then eat it up and destroy it—just like sugar-coating a tablet to make it attractive to a child.

Immunoglobulin 'E' (IgE) antibodies—the antibodies which atopics make so much of—become attached to yet another group of cells, called 'mast' cells. Mast cells contain substances that cause inflammation. Inflammation is rather like an explosion, and mast cells are rather like bombs. The IgE antibody is like a trip-wire attached to a bomb; when disturbed it makes the bomb explode. When a foreign protein meets the right IgE antibody attached to a mast cell, it reacts with it and thereby sets off the explosion. The mast cell releases its contents and these cause inflammation.

The body has complicated ways of controlling the amounts of antibodies it manufactures. In atopics there seems to be a fault in the control system and, as I have already described, an abnormally large amount of the IgE type of antibody is made.

Antibodies are made against foreign proteins from *all* sources. Foreign proteins are also present in foods, in animals, and in plants as well as in viruses, bacteria, moulds, and parasites. Although these substances do not actively harm us—unlike the viruses and so on—it is still important to keep them out of our bodies, because the proteins they contain could cause mischief simply by getting in the way.

The atopic state, *atopy*, is inherited; it is passed on from parents to their children. It isn't so much the diseases themselves—atopic eczema, asthma, and hayfever—that are inherited, as the predisposition to develop them. This is the reason why parents who have none of these diseases themselves can unexpectedly have a child who does; it is probable that at least one of them is atopic without ever realizing it. This is also the reason why parents who had atopic eczema themselves may have children who never get either eczema or asthma or hayfever.

In summary, then, an atopic person is someone who is predisposed to develop certain diseases—the atopic diseases of asthma, eczema, and hayfever. If an atopic person does not develop any of these diseases, he or she may be perfectly healthy throughout life despite producing large amounts of immunoglobulin 'E' antibodies. Measuring these antibodies is a way of detecting atopic people. Healthy or not, atopics may pass the state of atopy on to their children, who in their own turn may or may not develop these diseases. We will be talking about inheritance and atopic eczema in more detail later.

These concepts are probably the most difficult I will attempt to tackle in this book. If you have been able to follow the discussion so far, the rest will be easy.

4

What is atopic eczema?

Atopic eczema is largely a disease of children, and a very common one. Perhaps as many as 15 per cent (i.e. one in seven) of all children have eczema at some time or another.

THE NATURAL HISTORY OF ATOPIC ECZEMA

Atopic eczema affects people of all ages. Nevertheless, it is predominantly a disease of children, and is especially a disease of very young children. Eighty per cent of all who develop it first do so before the age of one year (*Figure 4*). Curiously, it is rather unusual for it to make its initial appearance during the first month of life, a time when other rashes are very frequent. These other rashes include *miliaria* (so-called 'sweat rash'), *napkin eczema* ('nappy rash') (*Figure 5*), and something known as *seborrhoeic eczema* (*Figures* 6 and 7). These rashes may be

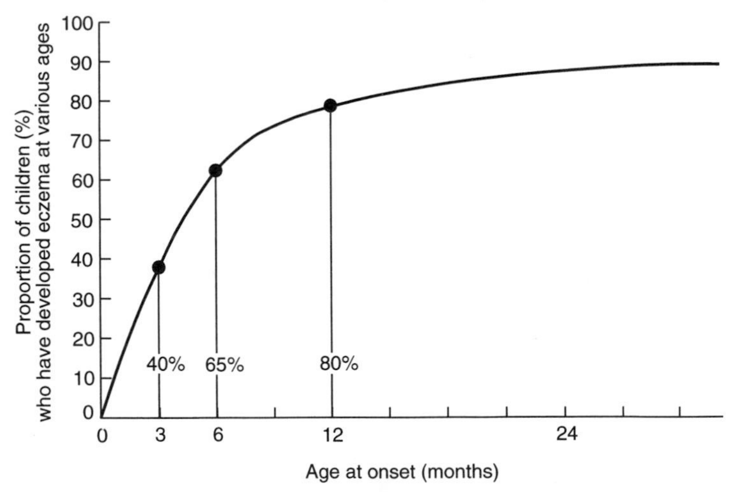

Figure 4 The age of onset of atopic eczema. You can see that the condition will have appeared by the age of three months in 40 per cent, and by the age of one year in 80 per cent.

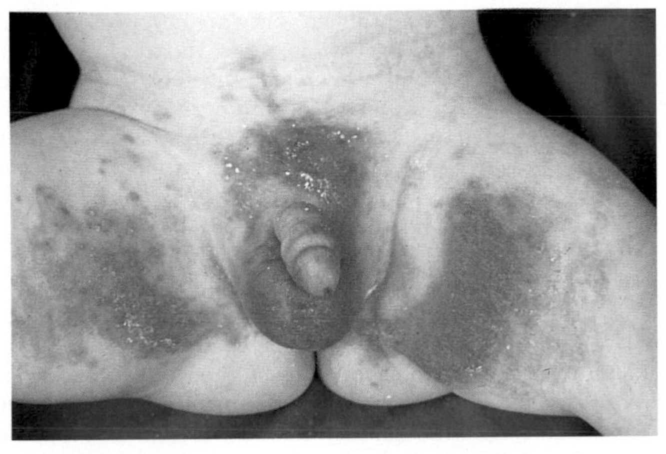

Figure 5 Typical irritant type of napkin dermatitis, showing sparing of the deeper folds of the groins.

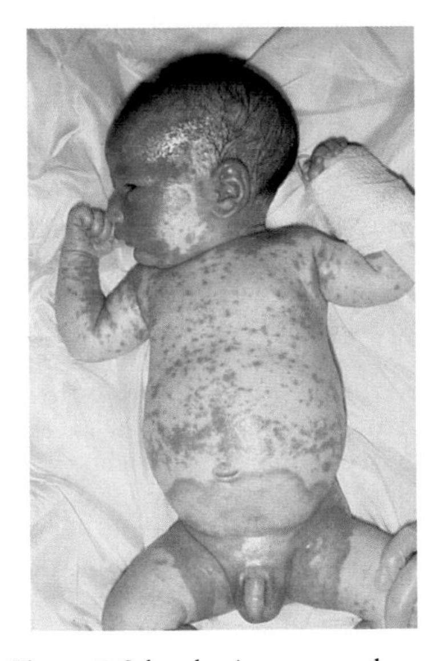

Figure 6 Seborrhoeic eczema, showing its typical distribution on the face and in the larger folds of the neck, armpits, and groins.

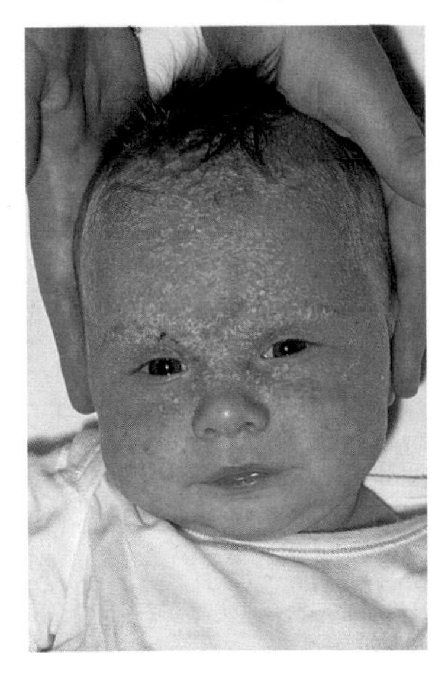

Figure 7 Seborrhoeic eczema, showing the common pattern on the scalp and upper part of the face.

trivial, or they may be quite a problem. They are nevertheless all fairly transient, and usually get better within a few months. Doctors may rather loosely call all such rashes eczema, and it is important to be aware that eczema does not always mean *atopic eczema.* Sometimes doctors use the general terms, 'eczema' or 'infantile eczema', when they are not certain whether they are dealing with atopic eczema rather than one of the other types of eczema that occur in babies.

Sometimes, one of these early rashes gradually transforms into more typical atopic eczema, and in retrospect it may seem likely that the rash actually was atopic eczema from the beginning. In fact, parents often tell me that their baby's skin was not right from birth, though it did not become clear that the baby had atopic eczema until a few weeks or months later. Very often, the skin has just appeared rather dry, and parents have frequently been told after the birth that the baby must have been overdue, even though they are fairly certain that this wasn't so. Whether these babies really have had atopic eczema from birth is an unanswered but fascinating question, particularly in terms of gaining an understanding of the causes of the disease, as we will see later in the book.

Napkin eczema can sometimes gradually spread beyond the nappy area and evolve into typical atopic eczema. In fact, it may be true that babies who have nappy rash more often go on to develop typical eczema than those who do not.

Nevertheless, atopic eczema does not usually develop in this way. Most commonly, the baby has good skin until the second or third month. The rash usually first starts to appear on the cheeks (*Figure 8*). Initially, it may not seem to irritate the baby very much, but increasingly it does so. As the baby starts to rub and scratch, the rash starts to worsen and may continue to worsen until the cheeks are raw and weeping. The rash may appear elsewhere, but the face tends to be the worst affected part of the body at this early stage. The sites where the eczema develops often reflect the way the baby sleeps, and exactly which parts of the body can be reached to rub and scratch. The cheeks themselves tend to be most severely affected in those babies who sleep face downwards, probably because their cheeks are wet and in contact with the sheet for most of the night. Small babies are often better at rubbing their skin on other objects than they are at scratching with their own fingers. A typical habit is for babies to rub their backs by a frantic wriggling motion when they are lying on their backs.

Atopic eczema that starts on the face may gradually settle down and disappear, or it may become more persistent. In this case, it will usually spread very gradually on to the body and on to the arms and legs (*Figure 9*). Eventually, it is particularly likely to settle in the creases of

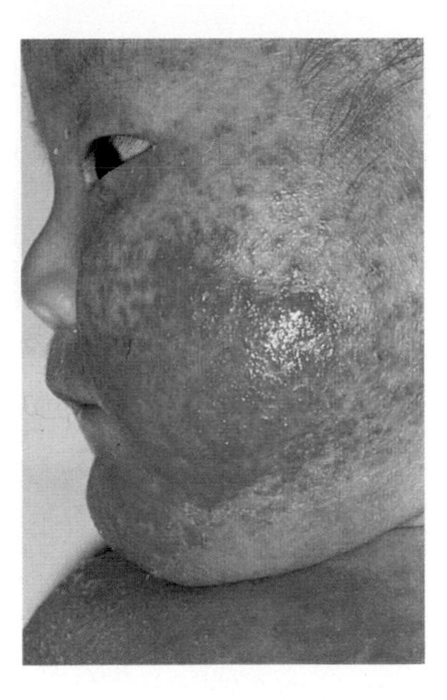

Figure 8 It is particularly characteristic for atopic eczema to start on the cheeks in babies.

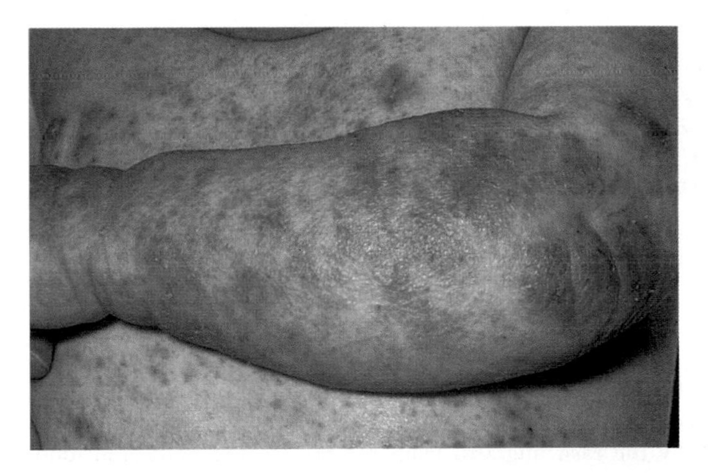

Figure 9 Atopic eczema tends to be widely distributed on the trunk and limbs in babies as well as on the face. At this stage, it does not show any particular predilection for the creases.

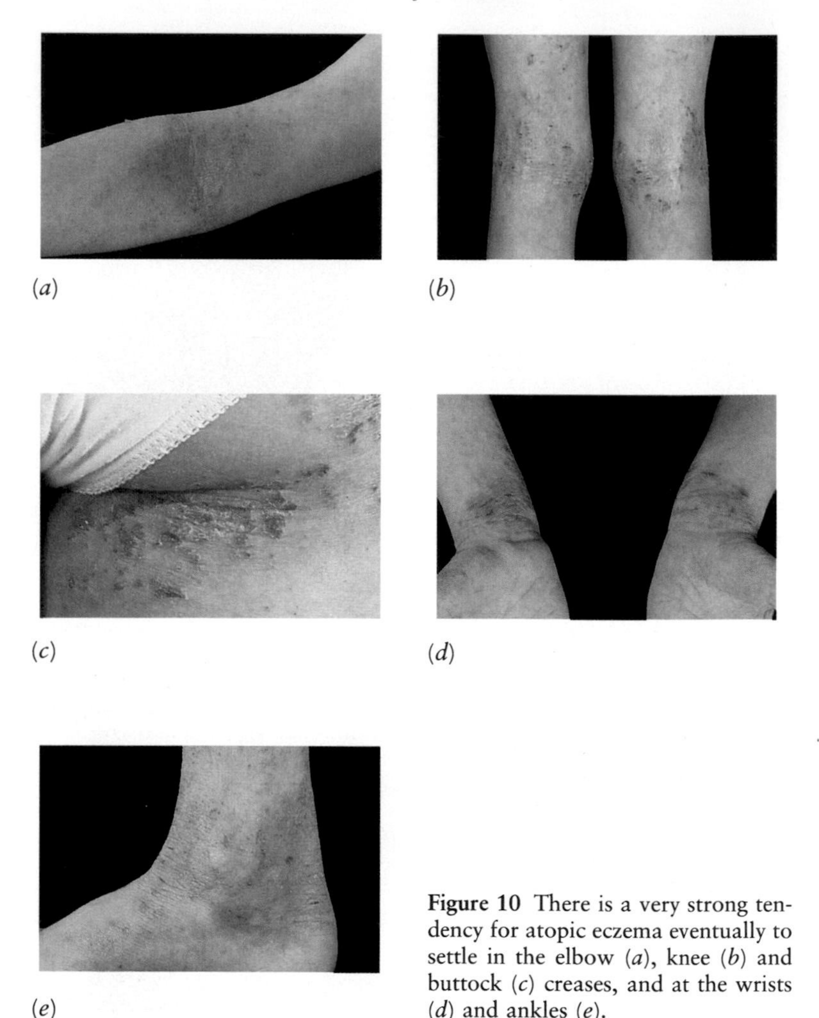

(*a*) (*b*) (*c*) (*d*) (*e*)

Figure 10 There is a very strong tendency for atopic eczema eventually to settle in the elbow (*a*), knee (*b*) and buttock (*c*) creases, and at the wrists (*d*) and ankles (*e*).

the elbows, wrists, buttocks, knees, and ankles (*Figure 10a–e*), and it is also common for it to affect the hands (*Figure 11*).

Why the rash migrates from the face to the body and limbs is a mystery; this is, however, the normal pattern (*Figure 12*). However, this is a disease that affects individual children in very different ways. Almost any site can be affected (*Figure 13*). Some sites are affected in

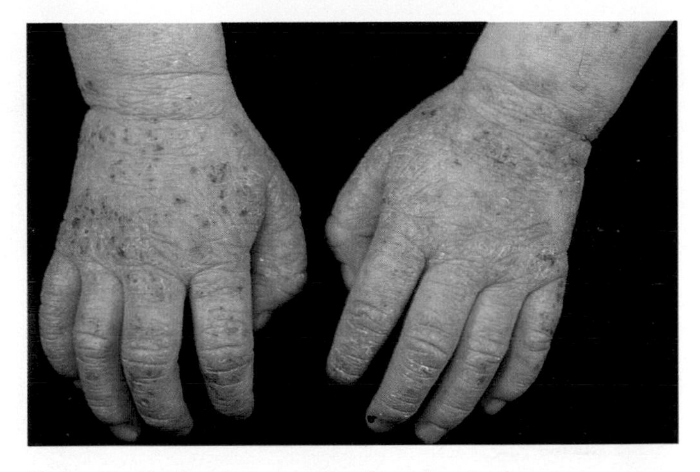

Figure 11 The hands are another favourite site for atopic eczema.

almost every child who has eczema, such as the crease in front of the ear lobe. Conversely, some sites, such as the skin on the nose, are very rarely ever affected. Occasionally a child shows a pattern in which the characteristic distribution in the folds is reversed, the rash being most prominent on the fronts of the knees and on the backs of the elbows (often termed the 'reversed pattern' of atopic eczema) (*Figure 14*).

In many, if not most cases, atopic eczema is fairly mild, no worse than a minor nuisance. However, in a few, it becomes severe, disfiguring, and disabling. In such cases, the disease seems to colour every aspect of the child's life, afflicting the parents with despair of an intensity rarely seen in any other childhood disease. In between these two extremes is the great bulk of children, whose eczema is of more intermediate severity.

One of the most distinctive traits of atopic eczema is a tendency to fluctuate widely in severity, often for no apparent reason. Later in this book, I will consider some of the factors that tend either to aggravate or to improve atopic eczema, though very often there seems to be no satisfactory explanation for sudden worsening or improvement in the condition. Of all the attributes of atopic eczema, the most characteristic of all is its unique behaviour in every affected individual. This is a disease that is never predictable and always eccentric.

Making the diagnosis of atopic eczema is generally a straightforward matter, but one or two unrelated conditions are occasionally confused

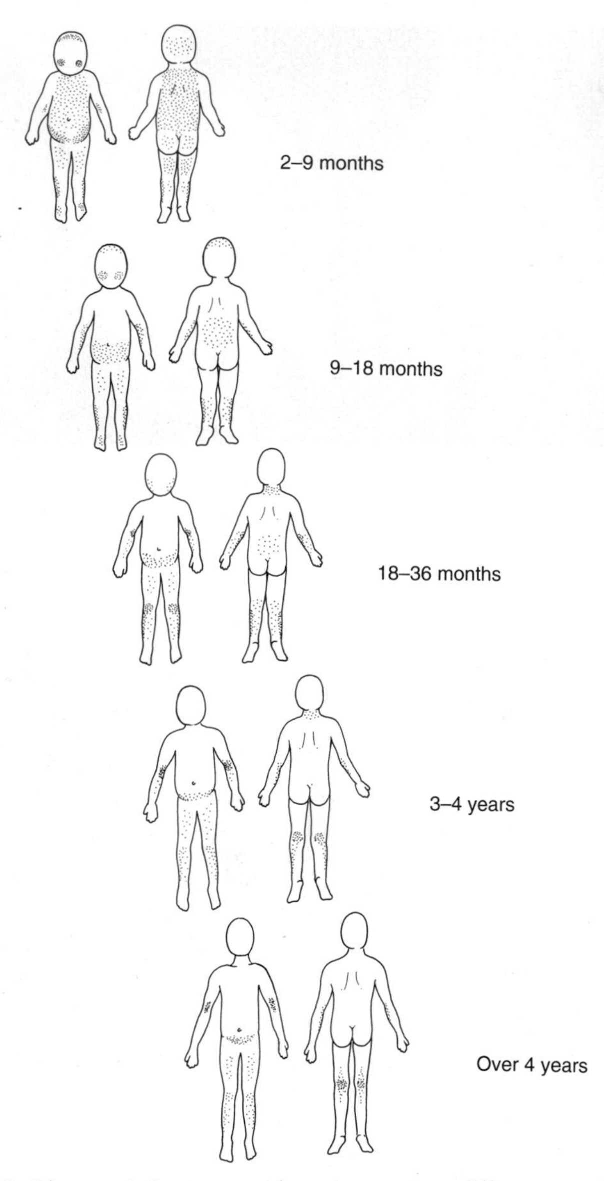

Figure 12 Characteristic patterns of atopic eczema at different ages. Why the distribution of the rash tends to change during childhood remains a mystery. Though these drawings show what most typically happens, in practice there is an almost infinite variety of patterns.

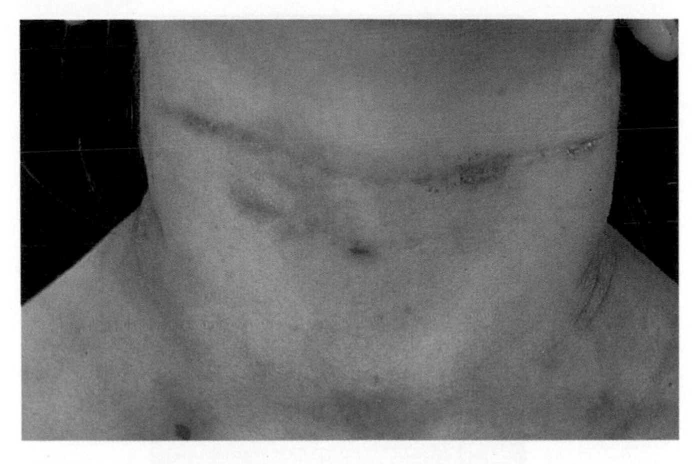

Figure 13 Atopic eczema may become a problem on the front of the neck, where it can be particularly aggravating. Very often, it results in fissures along the skin creases as in this child.

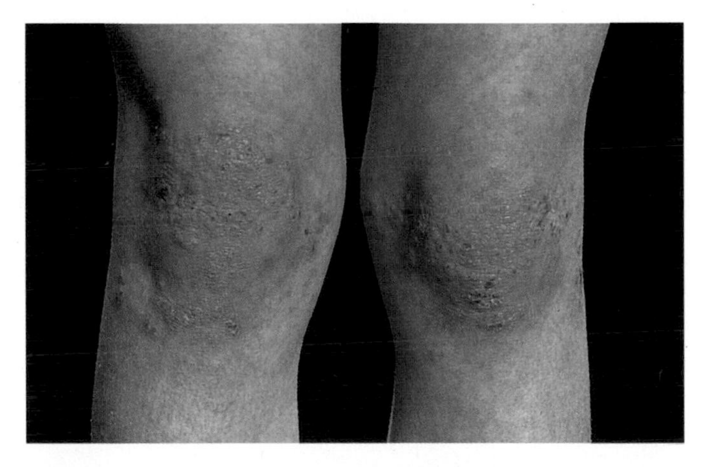

Figure 14 The 'reversed pattern' of atopic eczema on the front of the knees.

with atopic eczema. A condition that is particularly likely to be confused with it is a hereditary dry skin disorder known as *keratosis pilaris* (*Figure 15*). This condition takes the form of tiny raised red spots, very rough to the touch. The condition is generally most marked

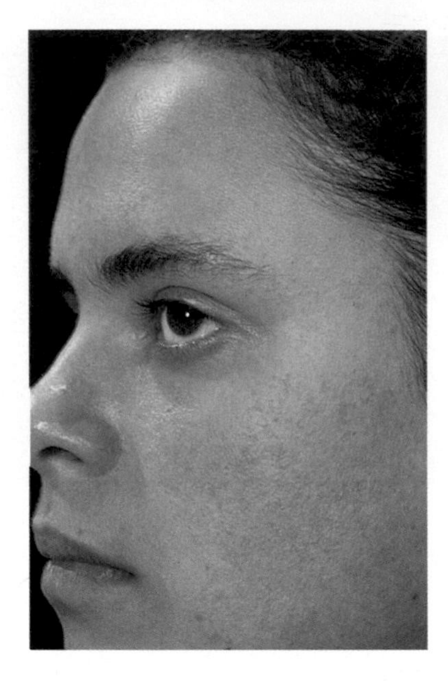

(*a*)

(*b*)

Figure 15 *Keratosis pilaris*, a condition that is often confused with atopic eczema, (*a*) on the cheek and (*b*) on outer upper arm.

on the cheeks, on the outer aspects of the upper arms, and on the front of the thighs. One can often feel the spots, which are only 1–2 mm across, better than one can see them. The redness may be very prominent.

ITCHING AND SCRATCHING

The hallmark of atopic eczema is the extreme irritation it causes. Few other skin diseases provoke quite such intense itching. In an attempt to relieve this itching, children will tear at living flesh. To the desperation of parents, there often seems to be no way they can persuade their child to abandon his apparent craving for self-destruction. The more effort that is made to do so, the more feverishly the child seems to claw his now bleeding skin. At times like this, the child may appear to be in an trance-like state, and may seem not to hear or respond to onlookers. One feels that were one to throw the child into the sea, he would sink to the bottom still totally engrossed in scratching. This is all devastating for parents.

Tiredness seems to make eczema more irritable, and undressing always aggravates it rather acutely; it therefore tends to be most troublesome in the evening and even more so, during the first hour or two after the child has been put to bed. In anyone who has eczema, psychological 'irritability' is always expressed as an increased itchiness of the skin. The mechanisms responsible for this are not known, but it is assumed that the brain's perception of the stimuli it is receiving all the time from the skin can change, depending on its overall state of arousal, so that it interprets these stimuli differently at different times.

Thus, when the brain is actively engaged in some very absorbing activity, itchiness can be more or less ignored, whereas at other times, when the individual is generally 'irritated', it registers a heightened perception of itchiness in the skin. It is no coincidence that the word 'irritable' is used both to denote a psychological state, and a sensation in the skin, as the two so commonly occur simultaneously, both in those with otherwise normal skin and, more intensely, in those with eczema or other itchy skin conditions. Similarly, when the brain finds itself relatively unoccupied, as it does when one is attempting to fall asleep, it does not suppress the itchy stimuli it is receiving from the skin, and they seem to become more conspicuous and troublesome.

Sudden cooling of the skin appears to be the reason for the instant increase in itchiness that follows undressing (see p. 63). Even those who don't have eczema will often have a good scratch when they get undressed, or when they get out of bed in the morning. This reaction to sudden skin cooling appears to be very much more pronounced in those with eczema.

Parents often say to me, 'If only you could do something about the itching, we could live with the rash'. Unfortunately, the two are inseparable. If one could deal with the itching, the rash would be beaten.

There is a school of thought that atopic eczema is merely the natural consequence of having a skin that is inherently exceptionally itchy. The theory is that everyone has skin with a unique and personal threshold for itching. Some people are therefore naturally more itchy than others —that is, their itch 'threshold' is lower. Those who believe that atopic eczema simply results from an unusually low natural itch threshold argue that the result is that scratching follows even the slightest stimulus, and that it is this scratching that induces the rash. According to this theory, putting a stop to the itching is all that one needs to do to get rid of the eczema. Disciples of the theory cite the observation that has occasionally been made that eczema seems to disappear when a child cannot scratch—for example, when comatose after road accidents, or sometimes when a limb is encased in plaster.

Personally, I cannot believe that this is all there is to it. However much a normal person scratches, this will not produce eczema, though it will eventually produce a change in the skin called *lichenification* (*Figure 16*), something which I will be considering later (see p. 28). Furthermore, most parents observe that sudden worsening of their

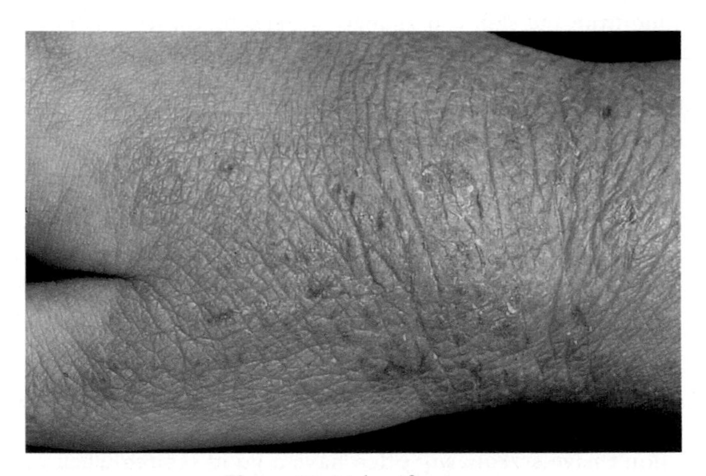

Figure 16 Lichenification.

child's eczema is preceded by some visible change in the skin that heralds the increased itchiness, though in practice the one tends to follow the other very rapidly. This 'herald' rash may be no more than a reddening of the skin, but sometimes there is also slight swelling of the skin in the affected area, which may be smooth but is occasionally 'pebbly'. Nevertheless, there can be no doubt that the damage done by scratching plays a major part in producing the rash we recognize as eczema, even if it isn't the only reason for it.

In reality, the itching in eczema is probably just one effect of the reaction occurring in the skin which is the real precursor of atopic eczema. During this reaction—which I will call the *eczema reaction* —certain chemicals are released within the skin which generate the sensation of itching. These chemicals probably do this by stimulating specialized nerve endings which then relay the sensation of itch to the brain. The sensation of increased itching is appreciated very rapidly after the onset of the eczema reaction, generally *before* the rash worsens visibly. An increased rate of scratching therefore tends to precede an obvious worsening of the rash, but this should not be misinterpreted as meaning that the scratching has *caused* the rash. Both are results of the same harmful reaction in the skin, the 'eczema reaction'.

REDNESS

The most regular visible characteristic of the rash of atopic eczema is redness, for which the medical word is *erythema* (another Greek word). In some children, the rash never really goes beyond redness with added scratch marks. In medical terminology, scratch marks are called *excoriations*.

When atopic eczema worsens rapidly, the first visible sign of this will almost always be intensified redness. Redness in the skin results from an increase in the flow of blood passing through the dermis (see p. 3). The rate of blood flow largely depends on the calibre (the internal diameter) of the blood vessels through which it is flowing. Increased flow results from an increase in blood vessel calibre (*vasodilatation*). Blood vessel calibre is altered by chemicals which reach the vessels either in the blood itself, or locally from the surrounding tissues. The source of the chemicals responsible for vasodilatation in atopic eczema remains unknown, although it may turn out that they both arrive in the blood *and* are produced locally. My own prejudice is that antigenic proteins from outside the body manage either to get into the blood or directly into the skin from its surface. Having arrived in the skin by either route, they are then recognized as foreign, and thus initiate the eczema reaction. The chemicals released during this reaction would have two effects; firstly they would cause itch, and secondly they would increase the calibre of the blood vessels, causing redness.

VESICULATION AND CRUSTING

The chemicals that cause the skin blood vessels to dilate also make the walls of these blood vessels 'leaky'. As a result of this 'leakiness', fluid seeps out into the skin from the blood. The effect of this is, firstly, swelling (called *oedema*), and, as the fluid penetrates into the epi-ermis, a bubbly, boiling appearance known as *vesiculation.*

Vesicles are fluid-filled blisters, which look rather like sago grains just under the surface of the skin. They are regarded as the most characteristic of all the features of eczema, and the word *eczema* actually comes from the Greek word for 'boiling'. Generally, these vesicles are tiny and not easily seen. However, very occasionally they may be larger and more prominent, especially on the palms or soles (*Figure 17*). When these vesicles reach the surface, they are easily broken by scratching off their tops. The fluid then oozes out on to the surface of the skin, and this is what makes eczema weep (*Figure 18*). The medical term for weeping is *exudation*. Because the fluid contains clotting substances from the blood, it may sometimes coagulate to

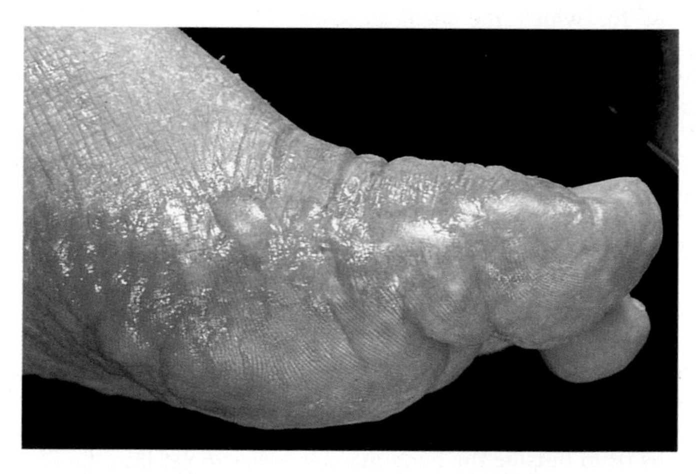

Figure 17 Very large vesicles on the foot.

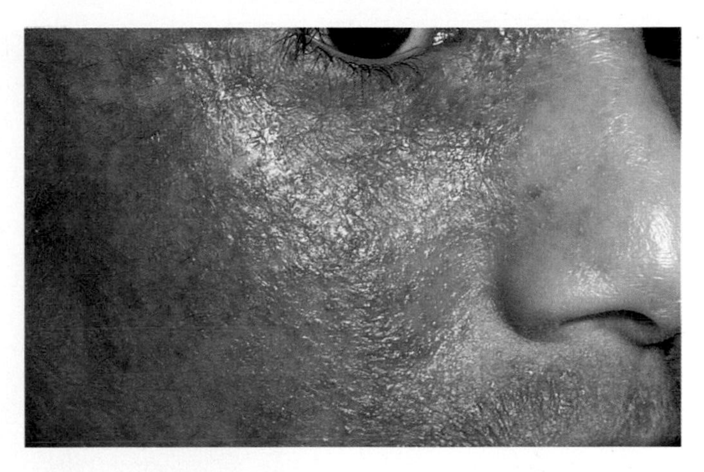

Figure 18 Weeping eczema on the cheek.

form *crusts*, more popularly known as 'scabs'. The reason why the fluid that oozes out is usually clear rather than bloody is that the red cells in the blood (which give it its colour) are too big themselves to escape from the blood vessels. The increased leakiness of the blood vessel walls makes them like a sieve that lets fluid through but largely keeps particulate matter, like cells, inside. If the blood vessels are more seriously damaged, by scratching for instance, then red blood cells may escape, because there is now a hole in the sieve.

SCALINESS

Scaliness of the skin is another prominent feature of eczema. Of course, scaliness is common to many different skin diseases. As we discussed in the introductory chapter, the most important function of the epidermis is to produce the stratum corneum, a waxy protective layer just like the finish on good furniture. Normally, one is hardly aware of this stratum corneum as it is more or less invisible, even though it is being continually shed and renewed. If at the end of the day you shake out your stockings or socks, the white powder that emerges is the shed waxy material of the stratum corneum. But, just as the finish on furniture can crack and peel when it is damaged, so can this surface layer of the skin. Scaling occurs when the normal adhesive properties of the stratum corneum are impaired, and this happens in a wide variety of diseases.

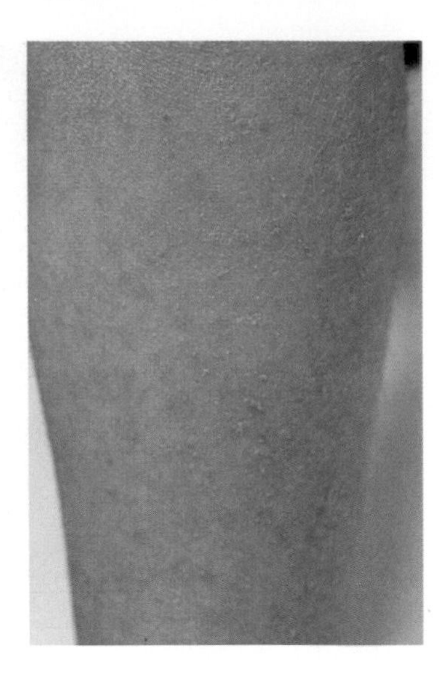

Figure 19 The dry scaly type of eczema.

Children with eczema often seem to have dry scaly skin (*Figure 19*). Some of this dryness and scaliness is a consequence of the disease. Even in areas where the other features of eczema are not apparent, such as redness and vesiculation, the eczema reaction tends nevertheless to be occurring to a mild degree, deep within the skin. This can do damage sufficient to cause scaliness without causing any of the other visible changes of eczema.

In some families in which someone has eczema, dryness of the skin may occur even in family members who do not themselves have any eczema. This sort of inherited dryness is called *ichthyosis* (from the Greek word *ichthyos*, a fish—because in some cases the scaling resembles the skin of a fish). Ichthyosis may also be found in families in which no-one at all has eczema. There are several types of ichthyosis which affect different families. In some families, the ichthyosis is of a severe type, but in others it may cause little problem. However, it is now clear that the commonest type of ichthyosis actually predisposes affected persons to develop eczema, though the exact reasons for this are not clear.

The medical term for this type of ichthyosis is *ichthyosis vulgaris* (which means common). In those who do not have eczema, it is usually manifest as rather mild, fine scaling seen most clearly on the arms and

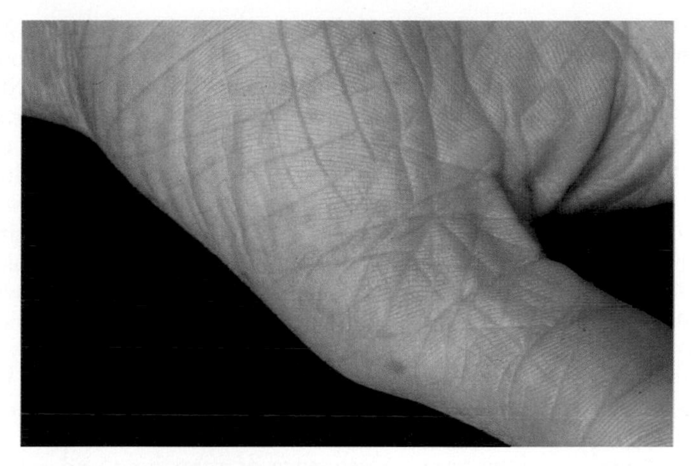

Figure 20 Palmar hyperlinearity.

legs. The severity and extent of this scaling is very variable, even between different members of the same family. Like other conditions inherited in the same pattern, ichthyosis vulgaris is passed from generation to generation through one parent, to about half of the children. A very characteristic feature of this type of ichthyosis is an excessively lined appearance of the palms, known medically as *palmar hyperlinearity*. The normal markings on the palms are exaggerated in number and degree, making the palms look prematurely aged (*Figure 20*). Ichthyosis vulgaris, whether or not it is associated with atopic eczema, is usually worse during the winter, and often improves markedly during adolescence.

Keratosis pilaris (*Figure 15*) has already been mentioned. It may occur on its own, in which case it is often mistaken for atopic eczema, or together, in which case it can be difficult to distinguish. Like ichthyosis vulgaris, this condition is always worse during the winter, and can be expected to improve considerably during adolescence. Ichthyosis vulgaris and keratosis pilaris quite frequently occur together.

PIGMENTATION PROBLEMS

Eczema is often associated with altered pigmentation in the skin. Skin pigment (known as *melanin*) is produced by cells in the epidermis known as *melanocytes* (see Fig. 1). People of all colours have the same number of these cells in their skin; the racial variation in skin colour is due to differences in the amounts of melanin pigment produced by these cells. Eczema disturbs the melanocytes, and can temporarily switch off skin pigment production. This effect can be prominent even where the eczema that provokes it is very mild. Fortunately this alteration of skin colour is almost always temporary, though it can last for many months even in areas of the body from which the eczema itself has disappeared. In children with a naturally light skin colour, loss of pigmentation may not be apparent except in the summer months when tanning occurs. Tanning fails to occur in patches where the eczema is active or where there has been eczema in recent months.

Another type of pigmentation problem can occur in children with a naturally darker skin colour, particularly those with black skin. In such children, darkening is associated with the thickening of the skin that occurs as a long-term consequence of scratching and rubbing (*lichenification*—see below). This darkening can cause considerable concern to parents of black or other racially pigmented children as well as to the children themselves.

In some older children who have had troublesome eczema around the neck for many years, this may be replaced by a curious 'dirty' pigmentation when the eczema finally gets better. Although this pigmentary change may remain for a long time, it will gradually improve and eventually disappear altogether.

There is a common misapprehension that these types of pigment change are due to the use of steroid creams and ointment. Though steroids can cause a variety of unwanted effects in the skin (see p. 112), they definitely *do not* cause these pigmentation problems.

LICHENIFICATION

An important characteristic of eczema is the change in the skin known as *lichenification* (*Figure 16*). This word is a very antique one and implies a resemblance to lichen, though this may not seem a very apt analogy!

If the skin is scratched and rubbed over long periods of time, the epidermis gradually thickens. A feature of this thickening is an exaggeration of the normal skin lines, and the overall appearance may come to resemble leather. This change is common in areas that have been affected by eczema for many years. Even if the eczema itself clears up, this thickening can last for months, and is itchy in its own right.

Some otherwise quite normal people have the habit of scratching a particular spot, as a sort of nervous release. The scalp, the back of the neck, and the shins are all a common focus for this kind of treatment, and eventually patches of lichenification will appear. These itch, and in this way a self-perpetuating cycle of scratching and itching can be set up. Much the same cycle is set up in eczematous children.

SCARRING

Although parents very understandably worry that their child's eczema will result in permanent scarring, this occurs extremely rarely. It never ceases to amaze me how dreadfully severe a child's eczema can be, and yet clear up leaving beautiful, normal skin, without a trace of scarring.

It may seem odd that a skin condition like chickenpox, which doesn't look any worse than eczema, leaves such marked scarring, and eczema doesn't. This is because chickenpox damages the skin at a much deeper level than atopic eczema. Although the 'eczema reaction' itself occurs at a deepish level, in the dermis, the skin damage one sees is in fact rather superficial, being mainly due to scratching.

As far as scarring is concerned, the critical thing appears to be whether or not the *basement membrane* (see Fig. 1), between the dermis and epidermis, is damaged. Once this membrane is damaged, scarring is likely. One sees this distinction quite clearly with burns; deep burns that damage the basement membrane cause permanent scarring, whereas more superficial burns, though they may be just as painful and look just as bad, usually do not.

Eczema herpeticum (see p. 80) sometimes causes some scarring. Similarly, areas of severe eczema where the skin is open and infected for long periods may leave slight scarring, though in most cases both of these types of scarring will eventually disappear. Loss of pigment should not be confused with scarring. As explained on p. 28, pigment loss is a complication of eczema that may last a few months, but tends not to last for years.

HAIR LOSS

Scratching is likely to lead to loss of hair. This may not be very noticeable on the open skin, but may be very prominent on the eyebrows and in the scalp. Fortunately the hair loss is only temporary since the hair follicles (from which the hairs grow) are not themselves damaged.

ECZEMA IN RINGS

Occasionally one sees eczema in the form of rings (*annular lesions*), most often on the trunk (*Figure 21*). These are very often mistaken for

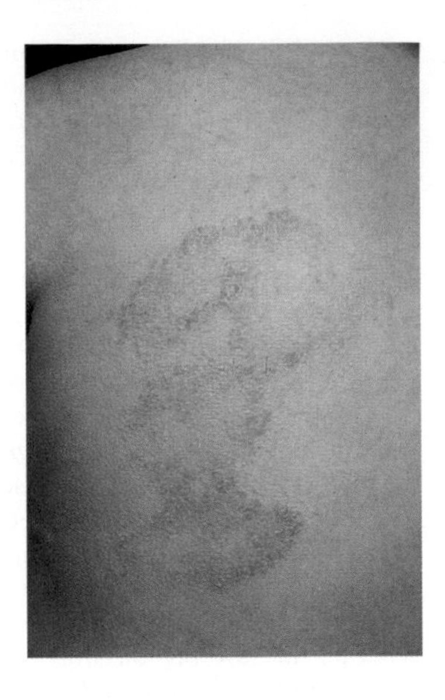

Figure 21 Ring-type eczema lesions on a boy's back.

the fungal infection known as *ringworm*. For this reason, children with this type of lesion may inappropriately be given treatments for fungal infection.

ECZEMA IN DIFFERENT ETHNIC GROUPS

Eczema affects all racial groups fairly equally, but the manifestations differ to some degree. For example, small Oriental and Asian children are particularly liable to develop swollen weepy coin-sized areas of eczema that can cause concern to parents and doctors alike, and which can be difficult to treat successfully (*Figure 22*). Disturbance of pigmentation is a greater problem in children with skin that is naturally more pigmented (*Figure 23*). Lichenification is always associated with darkening of the skin, which may cause a major cosmetic problem in those with darker skin (*Figure 24*). In many children, atopic eczema seems preferentially to affect the hair follicle mouths, the sites at which the hairs normally pass out of the skin. This causes an appearance of evenly spaced tiny spots looking very similar to keratosis pilaris (p. 19), but less rough, and often most prominent on the trunk. This change probably occurs in all ethnic groups but is visually most apparent in

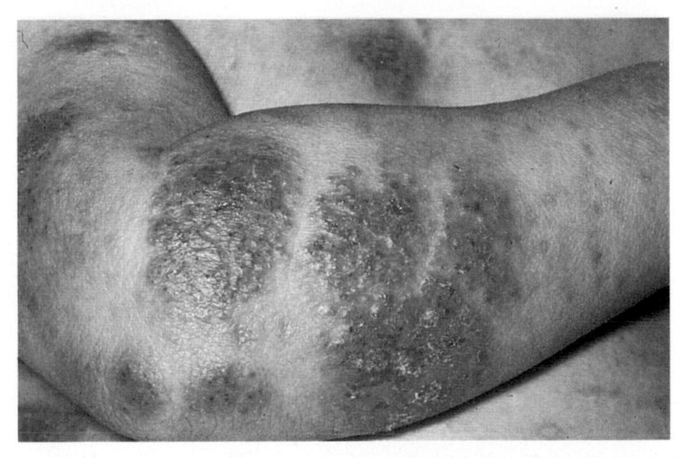

Figure 22 Swollen coin-shaped patches of eczema in an oriental child.

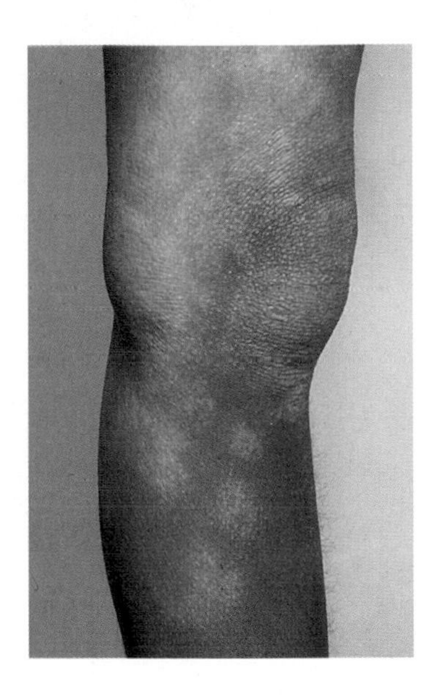

Figure 23 Disturbance of pigmentation in a black child.

Oriental and black children (*Figure 25*). The slight scarring that sometimes follows healing of areas that have been deeply scratched is much more apparent in pigmented skin (*Figure 26*). Such scarring can result in a substantial cosmetic problem in children whose eczema is other-

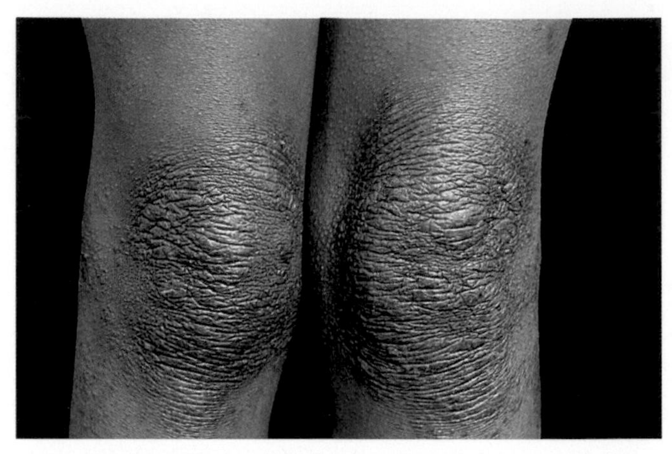

Figure 24 Lichenification of the fronts of the knees in a black child, showing the typical accompanying darkening of the skin.

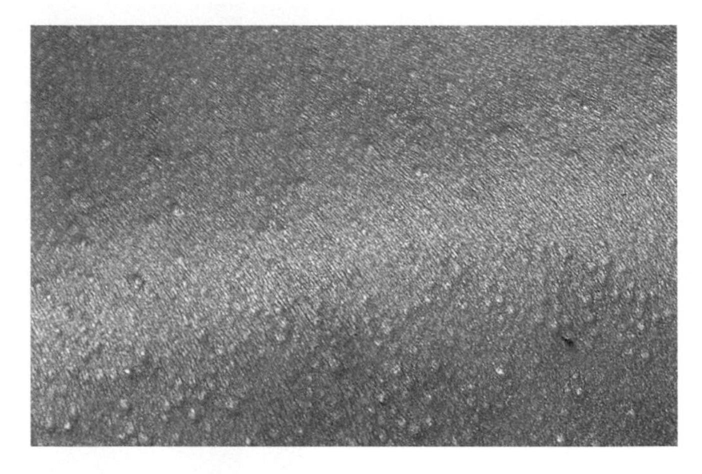

Figure 25 The 'follicular' type of eczema in a black child.

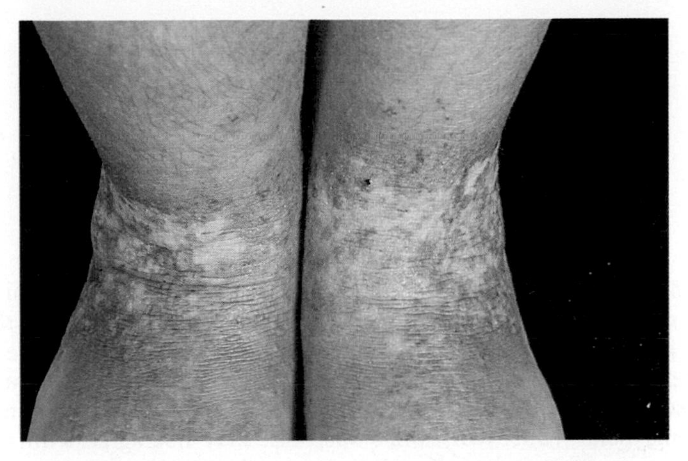

Figure 26 Scarring on the ankles in an Asian child.

wise improving. Fortunately, it is not permanent and very gradually disappears, though this may take several years.

SKIN INFECTION AND ATOPIC ECZEMA

The skin of those with atopic eczema can become infected by a variety of different types of micro-organism, including bacteria, fungi, and viruses. These infections contribute substantially to the problems experienced by those with eczema, but they are neither easy to recognize nor to treat. They are such an important aspect of atopic eczema that I will consider them in some detail in a separate chapter (p. 72).

5

What are the causes of atopic eczema?

I should love to be able to tell you what causes atopic eczema, but no-one really knows. As if to compensate for this lack of knowledge, there are almost as many different opinions on the subject as there are experts. My own opinion is just one of these, but my views on the causes of atopic eczema are shared by many other researchers in the eczema field. Your own child's specialist may not agree with what follows, and his or her view may eventually prove to be the correct one.

THE ROLE OF ALLERGY

I believe that atopic eczema is a response to substances in the environment which are harmless to the great majority of people. This 'eczema response' appears to be generated by the body's immunological system, whose principal function is to provide the body with an effective defence against infections and infestations. In individuals with atopic eczema, the immunological system responds to harmless substances in an excessive and an inappropriate way. The response is 'bungled', and the eczema is the visible consequence of the inflammation that results.

The botched immunological response that causes eczema could be called the 'eczema response'. This type of inappropriate reaction of the immunological system to normally harmless substances is known as an *allergic reaction*. The word *allergy* is often used in a much broader sense, to include *any* adverse reaction to such substances, especially in the case of foods. This way of using the term *allergy* is incorrect, because not all harmful reactions to foods are allergic in the true sense of the term, as they are not all generated by the immunological system. Before applying the terms *allergy* or *allergic*, one *must* have convincing evidence that the reaction is provoked by the immunological system. To illustrate this principle, one should recall the sad death of the old king of the elephants in *The story of Babar* (*Figure 27*), after he ate some toadstools. He was *poisoned* by chemicals in the toadstools, nothing at all to do with the immunological system.

(*a*)

(*b*)

Figure 27 Not all unwanted reactions to foods are caused by allergies. From *The story of Babar* by Jean de Brunhoff (Librairie Hachette, Paris and Methuen Children's Books, London). Reproduced by kind permission of the publishers.

I can give another example that is perhaps more relevant to the child with eczema. Most parents of such children are aware that wool next to the skin will make their child's eczema worse. At one time it was thought that this was due to *allergy* to the wool. However, we now believe that wool fibres generally make eczema worse simply by *physically* irritating the skin. In other words, the aggravation of a child's eczema by wool is usually not an allergic reaction.

In Chapter 3, we considered the way the body's immunological system works. Its first job is surveillance; it must be able accurately to recognize foreign material that succeeds in entering the body by whatever route. Having done so, its second job is to deal with that intrusive foreign material in an appropriate way.

We can again use a national defence system as an analogy. The coastline is under constant surveillance by radar, by air force planes and by naval ships on patrol. Any plane that enters our airspace, or any ship that enters our territorial waters, is quickly detected and unless its identity can immediately be established, a plane or helicopter is dispatched to take a closer look. If, on closer inspection, it is decided that the intruder is a threat, it will be intercepted. At this point there are a number of options. If the threat were considered very slight, it would be regarded as appropriate simply to escort the intruder away. On the other hand, it could be escorted to a home airfield or port and detained, or if it were regarded as a major threat, it might be considered safest to destroy it. Again, under certain circumstances, the incident might be seen as requiring some even more aggressive response, such as the bombing of a foreign city.

In peacetime, the first of these alternative responses is usually considered to be the appropriate one, and intruders can in this way be dealt with quietly and without provoking escalating retaliation. If, however, tomorrow morning the air force responded to an intrusion by an off-course Air India airliner by dispatching a squadron of bombers to bomb Delhi, there would be a dreadful fuss. All over the world the reaction would be regarded as unjustified and excessively violent. This situation is rather analogous to an allergic reaction; it is similarly excessive in its level of violence, and equally inappropriate to the danger represented by the provoking stimulus.

I see atopic eczema as just such an ill-adapted response to substances encountered in everyday life that we would normally regard as completely harmless. Many people think that a person who is allergic will only react adversely to one or possibly two substances. However, the situation appears to be very different in the case of those who have atopic eczema. In contrast with the person who is allergic to a single item, such as someone who develops urticaria (see p. 159) after eating lobster, it seems that the eczema skin response can be provoked by an extraordinarily wide variety of stimuli in any one individual.

THE ROLE OF FOODS

Generally speaking, it is only proteins that are able to provoke the immunological system directly. When they do so, they are called *antigens* (see p. 8). The greatest assault on the body by foreign proteins is provided by the food we eat each day, and it is likely that foods are of special relevance to atopic eczema, if only for this reason. However, I do not think for a moment that foods can act as the provoking antigens in atopic eczema, and it is almost certain that many other protein-rich substances in our environment can also do so. Among these, household dust, pets, pollens, and moulds are likely to be of particular importance.

However, the situation is almost certainly even more complicated than this, because many of the stimuli that appear able to aggravate atopic eczema in the predisposed individual do not appear to do so by means of an allergic response in the true sense of the term. The example of the response to wool (see p. 41) illustrates this point. We know that other purely physical stimuli, such as detergents, temperature changes, and emotional stimuli such as frustration, can also aggravate atopic eczema, and these certainly do not do so by acting on the immunological system, at least not directly.

Nevertheless, atopic eczema is probably first and foremost a reflection of genuinely allergic responses to foods and other protein-rich substances encountered in everyday life. The spectrum of antigens that provoke atopic eczema is probably different in each affected individual, so that no general assumptions can be made in this respect.

I don't believe that anyone can have atopic eczema without being allergic to *something*. However, once a person has developed atopic eczema, whatever the relevant antigens, that person's eczema is then liable to be influenced in addition by a number of non-allergic stimuli of the sort we have considered above, i.e. woollen clothing, detergents, changes of temperature, emotional frustration, and so on. None the less, I do not believe that these non-allergic stimuli are able to cause atopic eczema on their own. Only allergic reactivity can do this. The non-allergic factors complicate and aggravate rather than cause atopic eczema.

At this point, let's take a closer look at the allergy question, and consider the way in which an allergic response to foods might trigger an eczematous reaction in the skin.

First we will consider what normally happens when we eat a meal. The food we eat contains a great mixture of molecules. These can be divided into two types, the ones that provide us with nourishment (*nutrition*), and the ones that do not. The former consist principally of proteins, carbohydrates, and fats, plus a few extras such as vitamins and metals like iron which we need in very small quantities. The latter largely consists of fibres in plant matter.

Digestion is the process by which the nourishing molecules are extracted from food by the body. The digestive process takes place in two stages. First, the proteins, carbohydrates, and fats are broken up into smaller and smaller sub-units, rather like reducing a house to a pile of bricks. In the second stage, these smaller sub-units are taken up into the blood through the wall of the intestines by a process known as *absorption*. Since it is proteins that are important from the point of view of allergic responses, let us focus our attention on what happens to them.

Still using the brick analogy, imagine you have been given a house which does not suit your needs. You want a house totally different from the one you have been given, built to your own personal design. So you demolish the house you have been given in order to get the bricks with which to build the house you want. Proteins, like houses, are made up of subunits, which are called *amino acids*. These amino acids pass across the intestinal wall into the blood. From the blood they can be taken up into cells all over the body and used as the building bricks to make a great variety of proteins to your body's own particular design.

The great majority of the protein molecules in each meal are broken down into amino acids in the intestines. However, a very small amount of protein escapes this process and makes its way intact into the bloodstream. Though the amount is minute, it is nevertheless enough to stimulate the immunological system. One way the immunological system reacts is by producing antibodies against food proteins that have managed to escape digestion. These antibodies attach themselves to the proteins against which they have been produced, and the resulting combination of protein antigen and antibody is known as an *antigen–antibody complex* or an *immune complex*. These antigen–antibody complexes are gobbled up by special cells called *macrophages*, which are present in large numbers in certain parts of the body, particularly in the liver, the spleen, and the lymph nodes. Once gobbled up by the macrophages, the complexes are dismantled into amino acids, just as should have happened in the intestines.

This sequence of events occurs every time we eat anything. It happens quietly, and without us being aware of it. It seems, however, that the eczema sufferer's body simply cannot do these things in the same efficient way. The amount of intact food protein entering the blood after meals appears to be greater in those with atopic eczema, and the reaction to at least some of these food proteins appears to be altogether a more violent one. The end result of this different, more disorderly response to intact food proteins that have managed to get into the body is the eczema itself. As I have pointed out before, this last conclusion is a personal one, but nevertheless a reasonable working hypothesis.

The person with atopic eczema makes antibodies against the food proteins they have absorbed, but these antibodies sometimes seem to be rather different from those made by ordinary people. When these different antibodies link up with the relevant proteins, they also form antigen–antibody complexes, but these complexes are not so attractive to the macrophages and they are not so efficiently removed from the blood. In consequence, they circulate in the blood in greater quantities, and it seems possible that when they reach the vast network of small blood vessels in the skin, they are able to trigger the chain of events that results in eczema.

It has been suggested that the reason why children get eczema in the first place is that their intestines are excessively permeable to intact and undigested food proteins, so that the normal system by which these are removed from the blood is overwhelmed. This increased permeability might occur in one of two ways. One possibility is that their intestines are naturally more permeable from the start, and that this greater permeability is the main reason that they are predisposed to develop eczema. The theory is that the very small amounts of food protein that normally enter the bloodstream in early life lead to development of the safe disposal mechanism we have discussed, whereas the larger amounts that would enter if the intestines were too permeable would overpower the system instead. The result would be a failure in these children to develop the normal mechanism for handling such proteins, and instead the encouragement of a harmful type of response in the form of eczema.

The alternative explanation is that the innate defect that predisposes to atopic eczema is a defect in the immunological system, whereby the normal amounts of intact protein that enter the blood in early life provoke an abnormal response. This abnormal response seems mainly to consist of production of the wrong kind of antibody. If this second explanation were the correct one, we would still need to explain the presence of increased permeability that is present later on. However, this might well result from an allergic reaction within the wall of the intestines that causes slight damage, leading to increased 'leakiness', so that larger molecules than usual can get through.

The second of these theories is the one that is currently more widely favoured, and there is quite an amount of evidence to support it.

CONTACT URTICARIA AND ATOPIC ECZEMA

Food antigens do not necessarily cause problems only when they are swallowed. Some people develop a type of allergic reaction in the skin

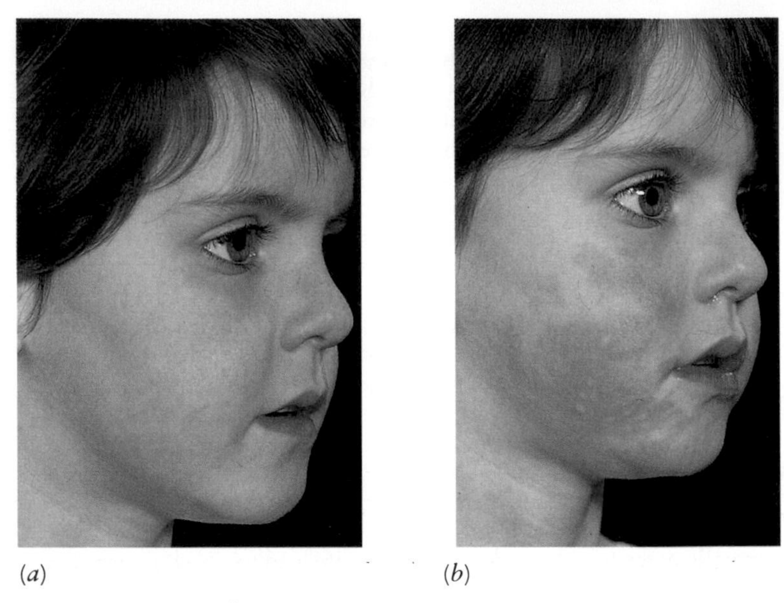
(*a*) (*b*)

Figure 28 Contact urticaria. This child developed intense redness and weals on the cheek within minutes of having direct skin contact with raw egg: (*a*) before; (*b*) after contact.

that is known as *contact urticaria*. This takes the form of redness, or, if the reaction is more intense, weals, at sites of direct contact with the relevant food, usually on the face around the mouth (*Figure 28*), or on the fingers and hands.

Contact urticaria occurs when the skin contains large amounts of IgE antibodies (see p. 10), and is exactly the same reaction that is elicited in the prick test (p. 8). The food antigen first crosses the skin, then reaches mast cells bearing antibodies of IgE type; the ensuing reaction between the antigen and antibody leads to discharge of the contents of the mast cells, and thereby to the development of inflammation of wealing type.

In order for this sequence to be initiated, the antigen must of course first cross the barrier provided by the skin. Normal, healthy skin will not allow the passage of antigens, and for this to occur the barrier must be broken. In those with eczema, the barrier will of course have been broken by the disease itself, and contact urticaria is a common event in affected individuals.

Another factor in determining whether contact urticaria occurs is the strength or *potency* of an antigen. Some antigens, such as those in egg white, in cow's milk, in nuts, and in fish seem 'stronger' than others, and therefore more likely to cause allergic reactions.

Therefore, for an individual child to get contact urticaria as a reaction to a particular food, it is necessary for that child to have:

- IgE antibodies being produced by the immunological system against antigens in that particular food, and bound in the skin to mast cells
- disruption of the skin barrier by eczema
- contact with a sufficient amount of the food
- sufficiently potent antigens within the food

Though common, contact urticaria is a type of reaction that is not invariably recognized by the patient, by parents, or by doctors; nor do they always realize that the reaction results from direct contact rather than from ingestion of the food.

Foods that provoke contact urticaria will not necessarily cause any problem once swallowed, but they nevertheless often do. Sometimes they provoke unpleasant symptoms following contact with the lips or with the inside of the mouth. There may be a tingling or burning sensation at these sites, or a nasty taste. The lips may swell, as may the lining of the mouth or throat. The latter can be very alarming to the child and parents alike, but fortunately such reactions are only very rarely dangerous (see p. 208). Following swallowing, a similar reaction may occur within the stomach or intestines, and may result in vomiting and/or abdominal pain. Subsequently, more generalized wealing of the skin (*urticaria*) may occur; this may depend on absorption of intact undigested antigen into the blood (see p. 38), and may take anything from a few minutes to about an hour after swallowing to develop. Absorption of food antigen into the blood may very occasionally cause a very serious allergic reaction called *anaphylaxis*, which is discussed in more detail in Chapter 9. In other cases, the food may aggravate the child's eczema, and this may happen either in addition to one or more of the above reactions, or in the absence of any of them. In reality, foods probably more often provoke eczema without also causing contact urticaria, generalized urticaria, or any other noticeable reaction.

The aggravation of atopic eczema by foods seems to take some time after the food is eaten, and it does not seem to occur every time the relevant food is eaten. The interval between the eating of the food and the aggravation of the eczema may be an hour or two, or a day or two.

Often the first thing that happens is that one notices a vague widespread blotchy redness of the skin, which occasionally resembles the rash of measles. This is accompanied by increasing itchiness. Following on from this redness, crops of tiny vesicles (p. 24) may occasionally develop. Such a reaction would be a relatively violent one; more characteristically, the whole affair is much subtler, and the eczema simply gets a little worse in the areas already affected, the worsening starting almost imperceptibly after a delay of an hour or two.

The foods that aggravate atopic eczema are most likely to be the types of food that are given to babies early in life and which remain prominent in a child's diet. Because these foods tend to be ones that are eaten many times each day, and because there may be several of them, the link between the eczema and the foods is only rarely an obvious one. In general, the severity of atopic eczema seems to fluctuate from hour to hour and from day to day, without any clear explanation for the fluctuations. This lack of an overt cause-and-effect relationship between foods and eczema is one of the main reasons that so many people, medical and otherwise, find it difficult to accept that there is any relationship at all. Nevertheless, we do now have fairly strong evidence that foods are a major contributory cause of eczema.

ATOPIC CONTACT ECZEMA

A great deal of research has been undertaken in recent years on the subject of allergic reactions in the skin in those with atopic eczema following direct skin contact with foods. As a result, it has become clear that contact urticarial reactions may be followed by the later appearance of eczema at the same site. This type of response has been called *atopic contact eczema*.

Depending upon several factors, particularly the intensity of the reaction and the ease with which the responsible protein antigen can cross the skin barrier, this reaction may occur within a few hours or up to three days after the provoking contact.

This type of reaction was first noticed in professional caterers who had hand eczema that was aggravated by direct contact with food, but it seems increasingly likely that it is important in other individuals with atopic eczema. Of course, most people do not have that much direct skin contact with foods except when they are babies, unless they are handling foods as part of their daily activities. This is particularly likely to happen with professional food handlers such as caterers, bakers, butchers, fishmongers, and those working in the food-processing industry, but may also be a problem in people preparing food at home.

After the early years of messy eating, children do not have that much direct skin contact with foods, though some will help their mothers to prepare food.

ARE ALLERGY TESTS VALUABLE?

Currently IgE antibodies to particular antigens (known as *specific* IgE) are detected by two principal types of test. The first type is a test to measure levels of these antibodies in blood, and the second is a test to measure the amount of antibody that has become attached to mast cells in the skin.

There are several similar tests which measure the levels of specific IgE antibodies in the blood in slightly different ways. All of them are done in special laboratories, and are rather expensive. The most widely used and most respected is the *RAST* (short for ***R**adio-**A**llergo**S**orbant **T**est*!). Another, newer test is known as *MAST-CLA* (short for ***M**ultiple **A**llergo**S**orbant **T**est-**C**hemi**L**uminescent **A**ssay*!).

The *skin prick test* is used to measure specific IgE antibodies attached to mast cells in the skin. A small amount of the appropriate antigen is introduced into the skin by a minute needle prick through a drop of solution containing the appropriate protein (*Figure 3*). In the USA and some other countries, the test is done by scratching the skin first, and then placing a drop of the solution on to the scratch (the *scratch test*). The reaction between the antigen and antibodies on the mast cells will lead to release by these cells of the substances they contain that provoke inflammation. This inflammation is manifest as a weal with a surrounding red halo. The weal reaches its maximum size by about 15 minutes, and its diameter is measured at this point, giving an indication of the concentration of specific IgE antibodies on mast cells in the skin. The reaction has usually died away by an hour. The test is not painful or more than slightly uncomfortable. Children are often frightened that they will be hurt by the needle, but, with the eyes closed, the tiny prick is virtually imperceptible.

Prick test reactions are in some ways more meaningful than the blood tests for specific IgE, because to some extent they mimic allergic reactions in real life. In practice, the test is considered positive if the weal diameter is 3 mm or more, and is generally regarded as negative if 2 mm or less.

There are, however, some snags with skin prick tests that mean they need to be interpreted with caution. For example, the size of the weal depends on the site at which the tests are performed, probably because the number of mast cells in the skin varies from site to site. The favoured place for skin tests is the inside forearm, where mast cell numbers are high.

Another factor to bear in mind is that skin test reactions will be wholly or partly abolished by antihistamine drugs taken by mouth in the previous 24 hours, and sometimes this effect may last as long as 48 hours. These drugs include Vallergan, Phenergan and Piriton (see p. 149). Other drugs do not appear to interfere, including topical and oral corticosteroids (see p. 106).

It can be particularly difficult to interpret skin test reactions in children with extensive eczema. If the test sites are affected by eczema, the mast cells seem to become exhausted and the weal size will be reduced or may even fail to appear at all.

Parents are often told that skin prick tests cannot be done in babies and very small children. Although it is true that test reactions tend to be rather less marked in the very young, they can be interpreted as long as the chance of a falsely negative reaction is borne in mind.

There are also several possible snags that concern the test solutions of which it is worth being aware. In this respect, the main snag concerns tests with foods. The test solutions are prepared using the raw food, for example egg, cow's milk, or wheat flour. However, we cook much of our food before eating it, and this will alter some of the proteins. Therefore, if an individual is allergic to one of these altered proteins, a skin prick test with the original uncooked protein may not be positive. Furthermore, as we have considered earlier in this chapter (p. 38), the proteins in foods are broken down in the intestines to smaller subunits called amino acids. During this process of digestion, molecules of intermediate size are produced, comprising several amino acids, but less than would be present in a protein; these are called *peptides*. It is likely, but unproven, that allergic reactions in the intestines themselves are sometimes provoked by these peptides rather than by the parent proteins. Like the parent proteins, these peptides are also absorbed into the blood in small quantities, and they may thereby lead to other types of allergic reaction, possibly including eczema. The problem with skin testing with food proteins therefore is that one is using solutions of the parent proteins, not the proteins that may result from cooking or the peptides that result from their digestion. The same problem will also occur with the other types of test for specific IgE, including the RAST, as these also use solutions of proteins exactly as they occur in raw foods.

Sometimes, a red, slightly tender swelling arises a few hours after a skin prick test, almost always at the site of a particularly large reaction. It is still not clear what this so-called *late reaction* means, but it is thought to indicate that the test protein is especially likely to be a provoker of genuine allergic reactions such as eczema and asthma. As these late reactions will almost certainly not be apparent until your child is back at home, it is a good idea to look out for them yourself, and to report them to the doctor who arranged for the tests to be performed.

However, by far the biggest problem with all these tests for specific IgE is to interpret them in a sensible way. A positive test implies that the test substance *could*, under appropriate circumstances, cause an allergic reaction in that individual. It is clear however that one can have specific IgE antibodies to a particular food, for example, but be able normally to eat the food without any problem. I have indicated some of the possible explanations for this paradox, but in truth we probably do not know all the reasons. Despite these *false positive* test results, one will generally find a positive test result using the appropriate test food in a child with contact urticaria. However, since these reactions are usually recognized by parents, it is questionable whether tests really contribute anything in most cases.

A negative test result does suggest that a food would be unlikely to cause contact urticaria or atopic contact eczema in an individual child. However, it certainly does not mean that the test food would be unable to provoke other types of allergic reactions, including atopic eczema, after it has been ingested. IgE antibodies do not seem to be involved in all such reactions, and the test is therefore inappropriate. An appropriate test for the identification of foods that can provoke atopic eczema following ingestion is simply not available currently, but would of course be of immense value if it were. The absence of such a test is the reason why so many unconventional tests are offered by private allergy clinics and laboratories, but there is no evidence that *any* of these provide results that are relevant. On the whole, it seems likely that most of the results obtained in such tests are merely spurious.

UNCONVENTIONAL ALLERGY TESTS

A wide variety of unconventional allergy tests are available, and I will attempt, to the best of my limited knowledge, to describe the principal ones.

Cytotoxic test

In this test, white blood cells from the patient are mixed with a solution containing the test food, and then examined under the microscope. The observer looks for abnormalities of white cell function, such as cessation of movement. The number of cells affected and the degree of the abnormality are together taken to indicate how harmful the food is likely to be to that patient.

When independently scrutinized, the results of this test as performed in most privately run laboratories have proved unreproducible. In other words, results differ when repeated on the same blood sample, and also differ from day to day in blood samples taken from the same individual.

However, though there are serious scientific doubts about its reliability, there is considerable interest in the principle of this test. This type of test has sometimes been criticized because the results tend not to agree with those obtained using the IgE RAST. However, the IgE RAST is a test for a very specific kind of allergy, and, as we have considered earlier, it is a test that does not seem to be good at telling us which if any foods are worsening a person's eczema. The criticism that the results of this test do not agree with those of the IgE RAST is inappropriate. It does seem sensible to look for tests that look at possible mechanisms that do not have anything to do with IgE.

Recently, interest has been revived by the development of an automated and more reliable version of this test, called the *ALCAT* test. In this form of the test, an increase in the size of white blood cells is sought following their exposure to the test food, and is measured automatically by a sophisticated machine called a cell sorter. The automated nature of the observations has made them much less subjective, and reproducibility appears much better. For these reasons, the ALCAT test warrants further research to establish whether it is better than the IgE RAST in identifying aggravating foods in the case of atopic eczema. For the present, there is no evidence either that it can or cannot do so.

Hair analysis test

Hair analysis is often used by 'alternative' practitioners to look for high levels of potentially toxic metals such as arsenic, lead, mercury, and cadmium, and for low levels of metals that are considered nutritionally important, such as manganese, selenium, and zinc. While it is possible, in good laboratories, to measure the levels of metals in hair fairly accurately, there are considerable doubts about the way that this type of test is used in privately run laboratories that offer the analysis to members of the public. There are anxieties about the reproducibility of the test in many of these laboratories, and on the validity of the interpretations and recommendations that follow from the results. First, hair can easily be contaminated directly; for example, medicated shampoos may contain selenium, so that results for this metal may say more about one's shampoo than one's health. Furthermore, we do not know, in the case of many of the metals, whether there is any relationship between hair levels and body levels, and for many of them we do not know what should be regarded as normal levels, even in the hair, nor which (if any) symptoms might result from their excess or deficiency. This type of test cannot be recommended, nor is it really an allergy test, and so perhaps it doesn't really belong in this discussion at all.

The methods used to detect food allergies using hair are various, but a common one is to swing a pendulum over the hair sample. 'Allergy' is demonstrated by an alteration in the characteristics of the swing of the pendulum. It is in truth very hard indeed to believe that allergy could be detected from a hair sample, and most of the methods used stretch credibility to the limit. Scientific surveys have been carried out in which samples of genuinely allergic and non-allergic individuals were sent to several of the laboratories that undertake such tests. It emerged that the results were totally unreliable in distinguishing between allergic and non-allergic individuals, and that the tests were as likely to fail to recognize that a person had a real allergy as one who did not, and often diagnosed allergies in individuals who were not allergic. Results on several hair samples from a single individual, which were labelled with a number of different names, varied dramatically. These surveys appear to have discredited this test rather firmly.

Vega test

This is a test in which a small electrical current is passed through the subject's body, in order to measure their electrical resistance, using a complicated piece of electrical equipment. Glass vials containing solutions of the test food are placed in the machine, and are believed to influence the result by emitting some kind of energy. This is another test that stretches credibility to the limit, and there is no evidence that its results are of any value whatsoever.

Kinesiology

In this fairly widely used test, the patient holds a stoppered glass bottle containing the test food in one hand, and a decrease in muscular power in the opposite arm is interpreted as evidence of allergy. The muscle power is judged by the patient's ability, for example, to resist a downward pressure on the outstretched arm by the examiner. Less frequently, drops of a solution containing the food are placed under the patient's tongue.

There is little scientific evidence available to allow comment on this technique, but what there is suggests that it is not reproducible, and it is therefore unlikely to be valid.

Neutralization–provocation testing

This technique is used in a variety of different ways by individual practitioners. Perhaps the most standard way is for potential allergens to be injected into the skin (generally using *intradermal* injections, which are deeper than the prick test). If either a symptom or a weal (or both) occurs with any of the allergens, either higher or lower doses of that allergen are then successively injected until the symptom or weal disappears. The last dose is called the *neutralizing* dose and is the dose chosen for desensitization treatment (see p. 180), usually by the sub-lingual technique.

There is no scientific evidence that this technique is valid, and some evidence that the provocation of symptoms in this way is more often a product of psychological suggestion than a result of the injections. Intradermal injection of allergens also carries the risk of provoking serious reactions including potentially fatal anaphylaxis (see p. 208), so that it should never be carried out in the absence of full resuscitation facilities.

CONCLUSIONS

From what I have been able to learn of these methods of allergy diagnosis, and from reading the results of scientific evaluations of their effectiveness, I would advise readers to steer clear of all of them for the present. Though no doubt some of the private laboratories that offer these tests do so in the honest belief that they have some validity, the test results obtained are rarely reproducible and usually meaningless. I suspect that quite often they are performed cynically and fraudulently, and I believe that there should be legislation to protect the general public from what is frequently a corrupt activity. I also believe that medical advice should only be given by those who have a recognized qualification to do so, and that both the performance of the tests and the giving of the advice should be scrutinized by an appropriate professional body.

You must decide, but I would advise that if you do have your child tested in any of these ways, you should take the result with a generous 'pinch of salt'. You should be aware that these tests are likely to miss a real allergy, and even more likely to detect 'allergies' which don't in reality exist.

NON-ALLERGIC SKIN REACTIONS TO FOODS

Confusion may be caused by the ability of some foods to aggravate eczema by mechanisms other than allergy. For example, parents frequently notice that tomatoes make the area around their child's mouth red. Other foods that tend to do this are oranges, other fruits, Marmite, and potato crisps. Such children may be genuinely allergic to these foods, but the redness is more often simply a consequence of direct irritation of the skin due to the acidity of tomatoes, oranges, and some other fruits, or the saltiness of the Marmite and the crisps.

THE ROLE OF NON-FOOD ALLERGENS

Food is very unlikely to be the only source of antigenic proteins that are able to provoke atopic eczema. Many other substances which are widespread in the environment contain proteins that almost certainly also play a role in causing or aggravating this disease. The most troublesome of these are likely to be house dust, pollens, moulds, and furry pets. However, while food proteins probably most characteristically provoke atopic eczema through the intestinal route and the bloodstream, the antigenic proteins in these environmental substances appear to harm those with eczema as a consequence of direct skin contact. These reactions appears to be identical with those I described above as *contact urticaria* and *atopic contact eczema*.

Let's take a closer look at these environmental substances that are now thought to be rather important in the aggravation of atopic eczema. I will consider pollens first.

Pollens

Pollens consist of individual cells which are required for the fertilization of seeds in plants; pollens are the equivalent of spermatozoa in animals (*Figure 29*). Pollen is transferred from plant to plant either by insects or by the wind. Those plants which use insect pollination generally have brightly coloured flowers, whose purpose is to attract the appropriate variety of insect. This type of plant generally produces only small quantities of pollen, which is sticky so that it adheres to the

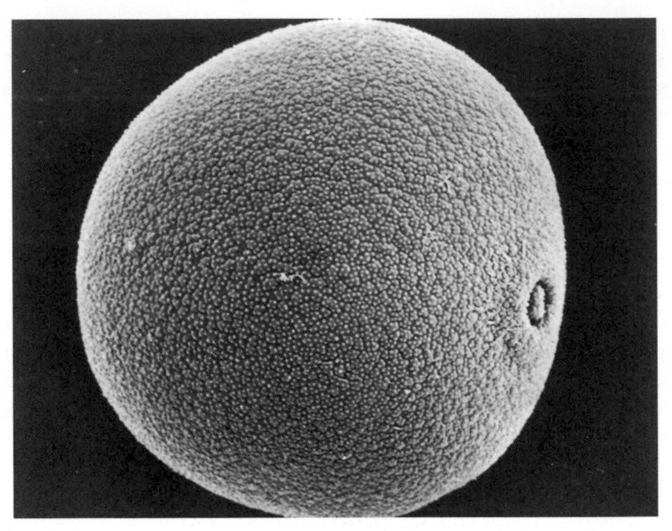

Figure 29 A grass pollen grain, magnified 2000 times. (Scanning electron micrograph reproduced by kind permission of SmithKline Beecham Pharmaceuticals.)

insect. This stickiness means that it does not become airborne in significant quantities. For this reason, the pollens of flowering plants only rarely cause allergies.

The pollen allergy problem is more usually caused by the pollens of those plants which depend instead upon the wind as a means of transport. Because this method is relatively inefficient in getting sufficient pollen to exactly the right spot in another plant of the same species, the pollen has to be produced in vast quantities to ensure that some will make it to the target. These plants tend to have small and unspectacular flowers, and the species which are most important from the allergy point of view differ from country to country. In the United Kingdom, the major offenders are grasses, followed in importance by a variety of trees, particularly alder, birch, hazel, oak, and beech. Certain plants with more conspicuous flowers are, however, largely wind-pollinated, and can therefore cause problems; the best known examples are mugwort (*Artemisia vulgaris*), oilseed rape, and nettles.

The production of pollen is a seasonal event. In the UK, grasses usually pollinate between the beginning of May and the end of July, and trees somewhat earlier—generally between the end of March and the end of May.

It can be difficult to establish whether pollen is an important aggravating factor in the individual. Tests for specific IgE antibodies can give helpful clues, but unless the test result is a very abnormal one, it certainly does not prove the case. In this type of allergy it is important to be aware that there are two important interacting factors, firstly the person's *sensitivity*, which is roughly measured by the skin test reaction or IgE test, and, secondly, that person's *degree of exposure*. One can be as sensitive as one likes, but without exposure, nothing will happen. On the other hand, massive exposure can lead to substantial problems in another person who is much less sensitive.

The most suggestive feature of pollen allergy is its seasonal nature. Thus, if your child's eczema is regularly worse during May and June, you may be justifiably suspicious of grass pollen.

House dust mites

As you might imagine, house dust contains a dreadful mixture of things. Although it has been known for many years that house dust can provoke allergic reactions, it was not known until recently exactly what it is in house dust that causes the problem.

It is now clear however that the main culprits are small mites, known by the general term *house dust mites*. There are in fact several different species of house dust mites, but the principal problem in the UK and in many other countries is a mite called *Dermatophagoides pteronyssinus*.

Dermatophagoides pteronyssinus mites live in our homes in vast numbers, literally millions (*Figure 30*). We are not usually aware of them because they are so small, about a third of a millimetre long, that they are invisible to the naked eye. Mites are technically *arachnids*, closely related to spiders, having eight legs (and are therefore not insects). For food they are dependent on finding organic debris of various types. Among the most important part of their food supply are shed skin flakes from humans and domestic pets. They appear particularly fond of some of the moulds which will grow on organic debris in suitably moist conditions. They are present throughout the year but their numbers tend to be greatest from August to October.

These mites are present in house dust where this accumulates in bedding, carpets, soft furniture, fabrics, and cuddly toys. House dust mites thrive in more humid environments, and are therefore happiest within three-dimensional structures such as pillows, mattresses, cushions, and cuddly toys. Moisture content tends to be higher in these

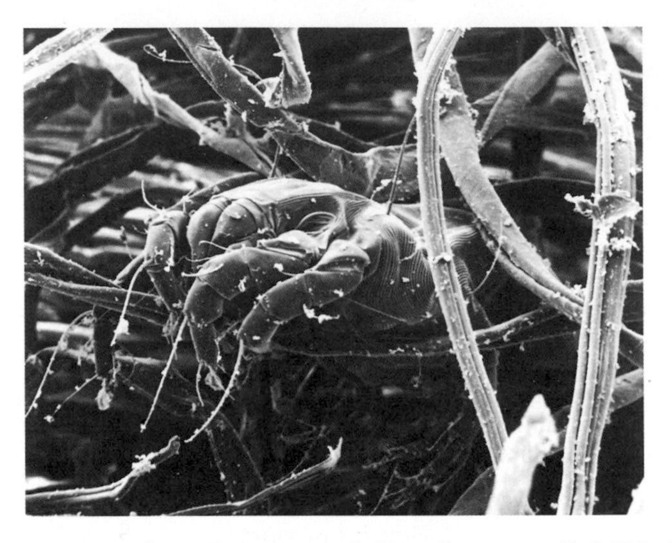

Figure 30 House dust mite amongst clothing fibres, magnified 225 times. (Scanning electron micrograph reproduced by kind permission of SmithKline Beecham Pharmaceuticals.)

situations than in room air, encouraging the moulds the mites favour as food as well as making life more comfortable for the mites themselves. The immediate environment of a child with eczema is likely to suit them ideally, because of the enormous supply of food from shed skin. It has been shown that the mattresses and bedrooms of eczematous children contain mite populations many times greater than normal.

Careful research has shown that it is not the mites themselves but their droppings that cause most of the allergy problem. Unlike the mites themselves, their droppings (or *faecal particles*) are a major component of house dust. These are approximately the same size as pollen grains, and can similarly become airborne under the right conditions.

Pets and other animals

Mammals are important sources of antigenic proteins, particularly where they are kept as pets in the home. Contrary to popular belief, it is not just their hair; in fact, their shed skin or 'dander' that is the greatest problem. Antigenic proteins are also present in their saliva and urine. Saliva proteins may be quite a problem as this can be left in the coat in considerable quantities in animals which spend a good deal of their time licking themselves. Some children will demonstrate their sensitivity very clearly by developing weals where they are licked. Similarly, traces of dried urine may be present on the coat as a result of splashing, especially in male dogs.

Nevertheless, dander is the main source of antigen, and cats and dogs shed remarkably large quantities of it into the home. The dander breaks up into small particles which can become airborne and will spread widely around the home, where they become part of the house dust. Like human skin, animal dander encourages house dust mites by providing an additional source of food.

Though the animals that are most troublesome to children with eczema are likely to be domestic pets, mammals encountered outside the home itself can also cause problems. Horses are particularly likely to do so. In the past, horses were probably the most important animals of all in terms of allergies, not only because there were so many of them around, but also because horse hair was widely used in fillings for furniture and bedding, and was even used in making the plaster on walls. Today, many children, especially those raised in the country, will have the opportunity of riding horses, and they will be encouraged to spend time caring for the horses by grooming them and cleaning out their stalls. This type of activity can cause a great deal of trouble, because the child is exposed not only to horse dander, but also to mould and urine in the bedding and hay.

It is difficult to ascertain whether there is much of an allergy problem with pet birds. Feathers themselves do not seem to provoke allergic reactions, but it seems likely that bird dander and droppings could. Feather pillows, quilts, and duvets do seem to cause problems however, but it is now thought that this mainly reflects the attractive environment that they provide for house dust mites.

Moulds

Moulds release vast quantities of spores into the air; these are the equivalent of their 'seeds'. This reaches its peak in the autumn months when there are a thousand mould spores in the air for every pollen grain. The main reason for this peak is the activity of moulds in rotting vegetation, especially fallen leaves. Compost heaps are a very rich source of mould spores throughout the summer and autumn. The principal outdoor moulds are called *Alternaria* and *Cladosporium*. Indoor plants may also be a source of these outdoor-type moulds. However, the main indoor-growing moulds are likely to be found on damp walls (especially *Aspergillus*), especially in cellars, and on stored food (especially *Penicillium*). Moulds also flourish in house dust.

Algae

Algae are similar to moulds, but are actually microscopic plants which live by photosynthesis. Like moulds, they are widespread outside but, in contrast, should be very little problem indoors. Like moulds, they often grow on the surface of plants, and lichen is in fact a combination of moulds and algae growing together. Algal spores are similarly dispersed in the air, but when they do contribute to eczema it is probably more often the result of direct contact with algae which grow on most surfaces outdoors, but particularly on other plants, such as grass. A child whose eczema appears to be aggravated by climbing trees may be allergic to algae that form part of the lichens on the bark. Children who itch after contact with grass may be reacting to algae (or moulds) rather than to pollen, and you should suspect that this is the case if grass seems to be troublesome outside the pollen season.

CONTACT URTICARIA, CONTACT ECZEMA, AND ENVIRONMENTAL ANTIGENS

I have described how foods can cause an allergic reaction known as contact eczema when they come into direct contact with the skin. I mentioned that this type of reaction, though probably very common, may not be very important in terms of provoking eczema in most patients, because, with important exceptions, few people get a great deal of food on to their skin. Most food passes fairly cleanly into the mouth, except in very young children, who are messy eaters, and who may use foods such as flour as play items. The other group who have a great deal of direct skin contact with food are of course those who spend time preparing it, both in the domestic situation, and more importantly, those who handle food professionally. In contrast, direct skin contact is an everyday event for everyone in the case of many of the important non-food antigens, like house dust mite droppings.

One of the principal questions addressed by researchers during the last few years has been whether skin contact with environmental antigens such as house dust, pollens, moulds, and animal dander is a cause of atopic eczema, even perhaps the principal cause. It is now considered very likely that it is a major contributory cause, but absolute proof is lacking and the relative importance of such contact compared to the role of ingested foods remains unclear.

When I considered food allergy earlier in this chapter, I explained that there is a condition called *atopic contact eczema* which can follow habitual food contact in a person who has IgE antibodies to a particular food. This contact eczema reaction may or may not be accompanied by obvious contact urticarial reactions to the same food, since this type of reaction is also a feature of having the same IgE antibodies. Clearly visible contact urticaria probably requires either heavy exposure or potent antigens. Where the antigen is weak or the exposure relatively light, it is probable that contact atopic eczema can occur without obvious preceding contact urticaria.

While contact urticaria is rather commonly provoked by contact with mammals such as dogs, cats, and horses, it seems to be rather a rare occurrence in the case of house dust, pollens, and moulds. This might be because exposure is lighter in the case of the last three, though frequently it may be more sustained. It might also be that the antigens in mammal dander and saliva are extremely potent.

It seems very likely that atopic contact eczema can also occur in the case of these non-food antigens. As in the case of foods, it may or may not be preceded by contact urticaria, but probably it is rather unusual for this to occur except in the case of contact with dogs, cats, and horses. It now seems likely that the constant low-intensity exposure we experience with house dust mite droppings, pollens, moulds, and other similar environmental substances more characteristically causes contact atopic eczema than contact urticaria.

This type of reaction may explain the worsening of eczema in some individuals during the grass pollen season, which is often most prominent at sites of contact with airborne pollen, i.e. the face, and at sites of direct contact with pollen still on the grass, i.e. the legs if the skin is exposed. But the greatest level of anxiety of all relates to house dust mite droppings. It would be very difficult to spot contact atopic eczema due to house dust mite droppings, because it is not seasonal, and because skin contact occurs every day and at most times of the day. Probably the most intense exposure occurs in bed because parts of the skin, particularly those not protected by nightclothes, are in direct contact with house dust mite droppings on the surface of the bedding, and, to make matters worse, penetration of the antigen through the skin will be encouraged by abrasion of the skin against the bedding, and of course by scratching. Similar intensity of contact will also occur on the legs and hands of children who play or sit on a carpet or settee.

Contact atopic eczema caused by non-food antigens like house dust mite droppings will tend to be most severe in areas of skin already affected by eczema. In these areas, increased moistness will serve both to trap the particles, and then to dissolve out the antigens which they contain. The skin surface barrier is also impaired to the greatest degree at these sites, so that these dissolved antigens will then be able to pass through more freely. The scratching, which is more concentrated in these areas, provokes even greater penetration of antigen. One can imagine that this could set up a vicious cycle.

As specific IgE antibodies play a part in contact urticaria and atopic contact dermatitis, tests for these antibodies using airborne antigens can sometimes be helpful. In fact, such tests are much more useful in the case of non-food antigens than they are in the case of foods. A strongly positive test to a pollen, to a pet, or to house dust mite does strongly suggest that these substances could worsen a child's eczema under appropriate conditions. The principal conditions that must be fulfilled are that contact must occur between an adequate quantity of the substance and damaged skin. It would be rather irrelevant, for example, to establish strong reactivity to cat dander in a child who never came into contact with cats.

THE ROLE OF MICRO-ORGANISMS

Some types of bacteria and yeasts can be found on the skin surface in people who have completely normal skin. The surface of healthy skin is rather inhospitable to such micro-organisms, and their numbers are therefore low. They appear to do no harm; on the contrary, these micro-organisms, known technically as the *normal flora* of the skin, probably do us good. Most bacteria and yeasts make antibiotic substances, which they secrete into their surroundings to reduce the competition. In this way, the normal flora probably provide us with some protection against harmful bacteria.

However, the skin of children with eczema is damaged, and provides a much more attractive environment for many types of micro-organism, bacteria, yeasts, and also viruses, many of which are able to damage the skin. We use the term *infection* to describe the situation in which they cause damage, which, in the case of a child with eczema, is superimposed on the damage caused by eczema itself. The subject of skin infections caused by these micro-organisms is dealt with in greater depth in Chapter 7. However, in recent years we have started to ask whether these micro-organisms might also play another, more subtle, role in eczema, apart from causing damaging infections. Bacteria, yeasts, and the waste substances and other products they fabricate contain antigens. It is presumably possible for a child with atopic eczema to develop an allergic response to these antigens just as easily as an allergic response is mounted to food and other antigens. The result would probably be a rather persistent type of contact urticaria and atopic contact eczema. Although it has not been proven that such allergic reactions to bacteria and yeasts actually occur in real life, it seems quite probable that they do. This occurrence of this type of reaction in some children with eczema would explain why antibiotic and antifungal drugs sometimes seem to be an extremely effective treatment, improving a child's eczema dramatically in addition to dealing with the infection for which they were really prescribed.

Recently, there has been a great deal of interest in the possible role that a yeast called *Pityrosporum* might play in atopic eczema. This yeast is part of the normal flora in adolescents, adults, and babies. It feeds on the oily skin secretion called sebum (p. 2). This secretion is dependent on sex hormone production, and is therefore present only at those times of life when such hormones are present in the blood. They are present briefly in babies before birth and for a short period after birth because they cross the placenta into the baby's blood from its mother. However, the resulting stimulation of the sebaceous glands lasts for many months. Sex hormones next appear in the blood when the child starts up its own production prior to the onset of puberty, any time from the age of seven years in girls, and about ten years in boys. Thereafter they remain in production throughout adult life, with gradually diminishing amounts being produced in later adult life.

As one might expect, *Pityrosporum* yeast populations are highest where sebum secretion is greatest, i.e. the scalp and face. Although this yeast is regarded as harmless by virtue of the fact that it does not generally cause infections, it does contain and produce antigenic proteins, and it now has been shown that these may provoke allergic responses in some individuals. Because the skin of those with eczema is broken and is therefore permeable to these antigens, and because those with eczema are atopic and therefore particularly prone to produce IgE antibodies, this type of allergy is almost exclusively seen in those with atopic eczema. The allergy seems to be both of the low-grade contact urticaria type and, more characteristically, of the atopic contact eczema type. It is therefore manifest as redness, itching, and worsening of the eczema on the scalp and face, and on the neck and upper trunk where dandruff falls, since this contains large amounts of the yeast and its waste products.

THE ROLE OF IRRITANTS

The skin of anyone who has atopic eczema tends to be unusually sensitive to a wide variety of irritants. This irritability will be a feature of their skin whether or not eczema is actually present at the time, and often continues long after the eczema has itself faded away. It is a more or less permanent phenomenon, and frequently of lifelong duration. It is rather similar to the irritability of the lungs observed in asthmatic children, in whom attacks of coughing or wheezing may be precipitated by chemicals such as sulphur dioxide (a major industrial air pollutant), paint fumes, and even fresh cold air.

The term *irritant* implies a directly harmful effect that has nothing directly to do with allergy. The irritants that aggravate atopic eczema can be considered under two principal categories: chemical agents and physical agents.

CHEMICAL IRRITANTS

A great variety of chemicals seem to be able to provoke or exacerbate atopic eczema after contact with the skin. On the whole, they are substances which would irritate the skin of perfectly normal individuals, but the eczema sufferer seems to be especially sensitive to them.

Soap, detergent, and water

Probably the most important chemicals of this kind are soaps and detergents, because of their widespread use. The function of both of these is to dislodge fats and oils by emulsifying them, and they are very effective at extracting natural oils from the skin. This process is harmful because the oils are there for their protective properties; their loss leaves the skin dry and vulnerable.

Soaps are generally milder than detergents, and many manufacturers add fats and oils to their soaps in an attempt to minimize this drying effect. Baby soaps are made with a particularly high content of fats and oils, and are therefore less likely to dry out the skin.

So-called 'housewife's hands' are hands dried out by detergents. In some people, this goes further than just dryness, and excessive detergent exposure is followed by eczema of their hands, even when they have never had skin problems before. However, the person who has had atopic eczema in the past is especially susceptible. The problem is just as bad, or even worse, in those who are exposed to detergents out of the home in their workplace. Such people include nurses, especially those who have to wash their hands frequently, and mechanics and engineers who have to use exceptionally powerful detergents to remove oily grime from their hands.

Paradoxically, water itself can have a drying effect. Even without the added effect of soap or detergent, it can leach out some of the salts and minerals in the skin and can remove some of its oils. Those whose hands are wet much of the time are prone to the same type of dryness and irritation as one sees in 'housewife's hands'; this can be a problem in such people as bartenders, fishmongers, fishermen, and butchers.

Children with eczema may find that contact between broken skin and water is painful, and they are especially likely to suffer from the misguided use of soap on their already inflamed skin. For these reasons, many parents avoid bathing their eczematous children. Many doctors also express the view that bathing is harmful in eczema and advise against it. I take the opposite view, that although bathing with water and soap, or even water alone, may be harmful, the bath is one of the most important elements in a comprehensive approach to treatment. Getting water into dry eczematous skin is one of the aims of treatment, but this water must be sealed in by oil. The bath can be used to deliver to the skin both this water and the oil seal. I will return to this subject in more detail when considering treatment (p. 96).

Many modern detergents for washing machines contain enzymes; these are often, but not always, described as 'biological'. These enzymes are mass-produced from cultured bacteria; they are natural substances whose normal function is to digest proteins and fats. They are very similar to enzymes we have in our own intestines for the same purposes. They have to be used at low temperature because they are inactivated by high temperatures, and they remove stains by digesting them away. There can be no doubt that enzyme-containing detergents are extremely effective. However, they are likely to prove extremely irritating if they come into contact with broken skin, as they will start to digest this in the same way. Healthy normal skin is protected from this type of irritant effect by the outer waxy stratum corneum (p. 2), but this protection is compromised in those with eczema. When clothes or bedding have been washed with such detergents, traces of enzymes remain adherent to the fibres. Here they can be reactivated by the moisture provided by the skin if the fibres come into direct contact with eczematous skin. Enzyme-containing washing detergents have proved one of the major domestic irritants of recent years, and should be carefully avoided by families in whom anyone has eczema.

Solvents

Solvents are fluids that dissolve other chemicals. The principal purpose of the solvents used both in the home and the workplace is to dissolve fats and oils. Skin contact with such solvents leads to the removal of protective fats and oils, a very harmful process—similar to the effect of detergents, but even more aggressive. Commonly used domestic solvents include dry-cleaning fluids, paint-thinners and strippers, petrol, paraffin, and aerosol propellants. Anyone with eczema should avoid skin contact with these completely if possible.

Some fruits and vegetables

Many fresh fruits and vegetables are highly acidic, particularly tomatoes and citrus fruits such as oranges, lemons, and grapefruit. Contact between the peel or flesh of these fruits and the skin may be highly irritating; many parents notice that their children get redness and itching around the mouth and on the fingers when handling or eating them. Most of the trouble caused by these foods is probably due to an irritant type of reaction rather than to actual allergy (see p. 34), though the two can be very difficult to distinguish.

Raw onions are another example of a food which is highly irritating to skin in its raw form, not so much because of acidity as because of certain chemicals it contains.

Salt

Salt is probably the usual explanation for the skin reaction that parents often notice around their child's mouth after eating Marmite or Bovril. Again, this is almost always an irritant reaction, though occasionally allergic reactions may occur with these items.

Antiseptics

Antiseptics are chemicals that kill micro-organisms, and it is hardly surprising that they can also be harmful to human cells. Some are highly irritating when they come in contact with skin, particularly in concentrated form. The still widely practised habit of adding domestic antiseptics to the bath will almost always be damaging to the skin of anyone prone to eczema.

Preservatives

Preservatives are in many ways very similar to antiseptics. They are chemicals added to foods, cosmetics, and skin treatments to prevent the growth of bacteria and fungi. Like antiseptics they are liable to irritate human cells to some extent, and some of them can occasionally cause allergic reactions. The application of a particular cream to a child with eczema may regularly provoke increased redness and itching. In children, this kind of reaction is due not usually to allergy but to the direct irritant effect of something in the cream, which probably only very rarely causes problems, and is most often a preservative. In this situation, dermatologists would regard the child as *intolerant* of the cream. Although true allergy to preservatives in creams does exist, it is rather rare in children. Remember that ointments less often contain preservatives than creams and lotions; this subject is considered in more depth later in this book (p. 92).

Smoke and other irritant vapours

Few people other than those who have the condition themselves are aware that smoke is very irritating to eczematous skin. In the domestic situation, this means cigarette smoke; no-one should be allowed to smoke in the presence of a child with eczema and, as far as possible, smoking at any time should be banned from rooms that the child uses regularly.

Other irritant vapours include those given off by dry-cleaning fluids, nail-varnish remover, spray perfumes, paints, aerosol deodorants, and 'air fresheners'.

PHYSICAL IRRITANTS

Fibres

The skin has contact with a wide variety of natural and man-made fibres in clothing, bedding, soft furnishings, and carpets. Pure wool tends to be particularly irritating to the skin of children with eczema. For many years, it was thought that they were allergic to it, but we are now fairly confident that the problem is caused not by allergy but by some physical property of wool fibres themselves. Children can certainly be allowed to wear woollen pullovers, but care must be taken to wear something underneath that prevents any contact between wool and the skin; even the small area of contact that may occur where a pullover overlaps at the wrists should be carefully eliminated. Pure nylon can have a similar harmful effect, and should therefore never be worn next to the skin; nor should nylon sheets be used on the child's bed.

Pure cotton seems to be the ideal material to wear next to the skin, but can be difficult to obtain, is relatively expensive, and its care is hard work. Cotton mixtures are the next best thing. It is much easier to obtain suitable clothing now than it was a few years ago, and the addresses of several specialist companies selling cotton clothing for those with eczema are listed at the end of this book.

Carpets of all types tend to be highly irritating and eczematous children should therefore play on them as little as possible. If necessary, cover part of the carpet with an old but clean cotton sheet.

Climatic factors

The interactions between climatic factors and the skin seem to be of great importance if one has eczema. Here I use the term 'climatic' in the broadest possible sense, as you will appreciate from what follows.

After years of listening to parents of children with eczema, I have the impression that the climatic factors that bother these children the most are extremely low humidity, conditions that induce sweating, and, perhaps most of all, *sudden changes* in temperature and humidity.

Humidity is the technical term for the amount of moisture in air. The main problem here is that very dry air tends to extract moisture from the skin, and therefore to make eczema worse. The relationship between humidity and air temperature is complicated and one needs to understand something of the science involved.

Warm air can carry more moisture than cold air. When warm air is cooled, a point is reached at which the moisture is forced out of the air and will start to become visible in the form of water. This is the basis of the formation of dew outside, and of condensation inside. The air on a hot summer day in England tends to contain a great deal of moisture, whereas the air on a very cold day in winter contains very little, even if the weather is wet. Below freezing point, the air will be exceedingly dry. When the air inside a house is heated up, the drying effect is increased because heated air tries to draw up more water than cold air. This means that humidity tends to be at its lowest in well-heated houses in winter, especially when it is dry or freezing outside. Increasing humidity with humidifiers or indoor plants can help, though it also suits house dust mites, and the benefits therefore have to be weighed against the disadvantages.

If reasonable anti-mite precautions have been taken, it is definitely worth considering the use of humidifiers in the child's room in the coldest part of the winter. The ideal is probably the type which heats water to a high temperature so that water vapour is released. However, these need to be controlled with a *hygrostat*, an instrument which measures air humidity and can switch the device on and off as appropriate. An address is provided at the back of the book where advice on humidifiers can be obtained.

Cold winter winds also seem particularly to aggravate eczema, but this is mostly because they increase the drying effect of cold air.

Eczematous skin seems able to adjust to different levels of humidity and air temperature, within limits. However, what seems to upset it most of all is sudden changes of either of these, which tend to occur together for the reasons explained above. Even those of us who do not have eczema feel a little itchy when we get undressed, when we get into a hot bath, or when we first get out of bed in the morning. Perhaps the best example is the effect of undressing, when the skin is suddenly robbed of the protective warmth and humidity of clothing. Undressing can precipitate serious bouts of scratching in children with eczema. The same effect is at work at night, every time any part of the child's skin becomes exposed to the room air after being enclosed in bedclothes or bedding. Whenever a child turns over in bed, some new part of the body is likely to be exposed; this starts the child scratching. Initially, the scratching may occur without waking, but soon the child wakes and things start to get worse. Unfortunately, turning over in bed occurs at precisely those parts of the sleep cycle where one is closest to being awake anyway.

Conditions that induce sweating are frequently a special problem. The sweat glands are located rather deeply in the skin (p. 2), and the sweat flows up to the surface in the sweat ducts. Eczema appears to result in damage to these sweat ducts as they pass through the superficial parts of the skin. As a result of this damage, the sweat is likely to leak out of the ducts into the living part of the skin, where it causes considerable irritation. A second result is the failure of the sweat to reach the surface, which means that cooling cannot occur. For both these reasons, children with extensive eczema can be very uncomfortable indeed in hot weather.

In some children, the eczema seems to be more or less confined to constantly exposed areas such as the face and hands, and one cannot avoid the feeling that some environmental factor must be responsible for this distinctive pattern. Though it seems likely that this is so, climatic factors are probably only part of the story. Sunlight is often blamed, but is in fact likely to be rather a rare provoker of eczema; more often, it is the radiant heat that accompanies the light. For this reason, open fires are frequently a problem. Other factors may play an equal or greater role in facial eczema, such as allergy to *Pityrosporum* yeasts (p. 57).

Nails

When considering physical factors that aggravate atopic eczema, it seems wrong to ignore fingernails, since they are probably the worst offender of all. Without scratching (*excoriation*), eczema would still exist, but it would be a pale shadow of the problem it represents when people really do scratch. When one looks at the skin of a child with active eczema, one sees a horrible blend of eczema itself and of the damage inflicted by scratching.

Very young babies scratch very little. They do sense the itching but seem to be unable to react by scratching. As the weeks pass, they begin to be able to do so. One of their first responses to eczema is to rub the affected part of the body on whatever is to hand, particularly bedding, carpets, furniture, and other people. They often perform quite gymnastic routines. Gradually they learn the use of hands and feet for this purpose, and initially the action is one of rubbing rather than scratching. When they do first learn to scratch, they are very clumsy and inaccurate. As a result, they can do a frightening amount of damage. This can be one of the most dreadful periods for parents, the more so because they are unable to communicate with their child and explain that this is counter-productive. The situation generally improves a little as scratching becomes more refined, and as the child begins to become aware of the link between reckless scratching and its painful consequences. It will often improve still further when parents are able to reason with their child, though this is a time when the situation starts to become complicated by the child's realization that there may actually be something to gain by scratching. This is a subject to which I will return later on (p. 227).

THE ROLE OF THE PSYCHE

The influence of the mind on atopic eczema cannot be underestimated. Psychological factors play a major role in both maintaining and aggravating the disease. Perhaps they can actually precipitate it in some cases. Conversely, psychological influences can also have a beneficial effect, an aspect to which I will return later (p. 188).

At one time, it was thought that children with eczema had a characteristic personality which was itself largely responsible for their developing the disease in the first place. More recently, it has been considered more probable that the personality traits commonly found in eczematous children are a consequence, rather than a cause, of the disease. In fact, there is almost certainly a degree of truth in both these views.

In children and adults with eczema, the skin acts as a 'thermometer' of emotional 'temperature'. Whereas, in mild cases, other children bite their nails or lips, or, in more severe cases, pull out their hair as a sign of tension, eczematous children take out their pent-up feelings on their skin. Stresses of whatever cause—being told off, being refused, being bullied, worrying about exams or about their place in a football team—are all directed on to the skin.

Observation suggests that eczematous children are generally more introspective and less extrovert than their non-eczematous counterparts. They seem more sensitive to the emotional stresses of everyday life, and this is reflected in a lowering of their threshold to anxiety. Put another way, a particular liability to anxiety or to feelings of insecurity does seem to underlie eczema in many children, even in babies. This probably contributes to the development of the eczema in the first place, and then to its maintenance. Nevertheless, it is almost certainly true that the rather characteristic personality traits of the eczematous child on the whole arise as a response to the disease. Persistent and unpleasant awareness of uncomfortable skin is highly distracting, making it that much more difficult for the child with eczema to take an interest in other people and in other pursuits. The child often seems to become obsessed with the skin, endlessly picking at it, making little piles of the bits that have been picked off or eating them. They become fascinated by the bleeding they cause, and it can be extraordinarily difficult to divert their attention from these activities. They become engrossed in scratching, sometimes to the point where it seems they behave as though drugged or frankly addicted to the activity of scratching.

From the emotional point of view, however, an even worse enemy than itching is shame. An eczematous child is not naturally ashamed of his skin, but may be made so by other people. Parents, brothers and sisters, grandparents, school teachers, and all sorts of other people can do untold emotional harm in this way. Once aware that other people find his skin repulsive, any pre-existing tendency to introversion and withdrawal becomes heightened. The stress caused by other people's attitudes, coupled with his own embarrassment, then becomes a potent force in the aggravation of the skin disease, and thus yet another vicious circle is established.

In conclusion, the psychological make-up of children and adults with eczema is simultaneously a result and a cause of their disease.

CONCLUSIONS

You can now see why the answer to the simple question, 'What causes atopic eczema?', is a very complex one. It is my own personal view that no-one develops the disease unless they are allergic to something in the first place, and that it is, at least initially, a manifestation of allergy. The 'something' to which they are allergic is almost certainly related to the age at which the eczema first develops. In the most common situation, where it starts in early infancy, something in the diet is probably the most usual source of the provoking antigen. In the infant who is bottle-fed cow's milk formula from the start, one of the antigens in cow's milk is likely to be the offender. Babies who are breast-fed from the start do also get eczema, if perhaps not so frequently, and in them the situation is much more complicated, as their mother's milk will contain many antigens that have originally come from foods she has eaten herself. In children who develop eczema later on, perhaps the 'contact' type of antigen is more likely to be the initial culprit—house dust mite, pollen, fungal spores, animal dander, or *Pityrosporum* yeast, for example.

However, once the eczema has got going, the situation seems to get more complex rather rapidly. Other allergies seem to develop in quick succession, some of which lead to aggravation of the eczema and some which do not. Thus, children with eczema rather characteristically react to a large number of antigens in skin tests or in the RAST, but it is not possible to tell which play a part in the eczema and which do not. The situation is further complicated by the role that is increasingly played by non-allergic factors, such as irritants of one type or another, and the psyche.

Complex interactions develop between these various elements. The condition called *exercise-induced, food-dependent anaphylaxis* (see p. 208) is a good example of the kind of interaction I have in mind. This is a disorder which causes athletic individuals to collapse in the middle of exercise, usually marathons, with urticaria, breathlessness, extreme weakness, and abdominal cramps. Although food allergy was suspected as a cause of these attacks, this explanation was initially rejected on the basis that they never occurred if the individual was not indulging in athletic exertion. Later on it became clear that the exertion was necessary for the allergic reaction to a food to develop. In other words, the food that was responsible would only cause the attack if eaten under very specific circumstances. This is a very important concept, ignorance of which is responsible for many misconceptions about allergic reactions among medical as well as non-medical people. It means that eating a particular food may not always cause the same reaction in someone who is allergic to that food, depending on the exact circumstances prevailing at the time. Exertion is almost certainly not the only factor that can modify allergic reactions in this way; such factors as the other foods that are eaten at the same time, what is drunk, whether one has an infection, and the state of the psyche, may all be relevant in this respect. The same will probably be true of non-food antigens. We can therefore conclude that other factors may modify those allergic reactions that contribute to atopic eczema, and both the allergic reactions and the modifying factors vary considerably from person to person and from occasion to occasion.

6

What is the outlook?

Childhood atopic eczema has a natural tendency to improve after a period. The great problem is to predict precisely when this is likely to happen in the individual case. What one can do, however, is to look at the statistics for large groups of children. Unfortunately, what emerges is that there is little agreement between the findings of different surveys. These inconsistencies seem largely to reflect the fact that in each survey the researchers were asking different questions in different ways. For example, a group of children known to have had eczema at the age of one can be re-examined, say ten years later. Some will appear still to have eczema; the eczema in others will appear to have cleared up. Even so, the chances are that some of those who are found to be free of eczema probably do continue to get it from time to time; they just happen to be clear when examined. This will tend to make the outlook seem more favourable than it actually is. Alternatively, one could send out questionnaires asking how the eczema is getting on. Several people will fail to reply, however, and the ones who do reply are more likely to be those who still have eczema. This will produce a bias in the other direction, so that the outlook will seem worse than it actually is.

In general, once it has appeared, atopic eczema exhibits three distinct types of variation in severity. The first is the most obvious, the typical day-to-day fluctuation which one can only rarely explain. The second is the tendency of the disease to show some degree of seasonal variation, being generally worse in the winter, spring, autumn, or summer, depending on the individual. The third is the almost inevitable worsening that follows the initial appearance of the disease, and the improvement that follows later in the great majority of cases.

It is the third type of fluctuation that we are most concerned with in this chapter. It generally takes a few years from its onset for atopic eczema to reach its worst level, but this is very variable. Of course, you will never know for certain that this level has been reached, until natural improvement is well under way. This natural or *spontaneous* improvement can be expected to occur in almost everyone with atopic eczema, whatever their age. However, it usually occurs so gradually as to be virtually imperceptible, until it is well under way. Improvement continues until the eczema is more or less clear, firstly for short periods of a day or two, then for longer periods until the skin is free of eczema for most of the time. The reappearances of the eczema tend to become less and less of a feature, but it is important to be aware that it is actually rather unusual for eczema ever to clear up completely and permanently, never to be seen again. In practice many, perhaps most, people who have had troublesome eczema in childhood will experience recurrences from time to time throughout life, though these may bother them very little. People who have had eczema as children always carry on having what many would call sensitive or irritable skin, which is made itchy by such things as soap, woollen fabrics, very cold weather, perfumes and certain other cosmetics, and often also by stress. For this reason, I prefer to avoid the expression to 'grow out of' eczema, as this seems to imply eventual total cure, which is not a realistic objective.

Because it is so difficult ever to say that a person's atopic eczema has gone for good, it seems to me that the appropriate question should not be, 'When will my child's eczema disappear?' but 'When will my child's eczema cease to be a problem?'. Despite the conflicting data on this, I think one can make some fairly sound generalizations. If a child has eczema at the age of one year, as most do, there is about a 50:50 chance that it will have stopped being a problem by the age of five, four years later. Every four years thereafter, half of the remaining children will see their eczema cease to be a problem. Therefore, of those children who have developed eczema by the age of one year, about 50 per cent will still be having problems at the age of 5, 25 per cent at the age of nine, 12 per cent at the age of 13, 6 per cent at the age of 17, and 3 per cent at the age of 21.

Figure 31 illustrates these statistics. As you can see, the curve never quite reaches the bottom of the graph. In other words, there are individuals who will continue to have eczema into adult life and, unfortunately, among these will be a few in whom it continues to be severe. Happily, such individuals constitute only a small minority of those who had eczema in early childhood, and even these individuals have a good chance of clearing up later on.

The picture is complicated by the fact that the outlook for anyone with eczema is influenced by the age at which the eczema first appears. In general, the later the onset, the less favourable the prospects of an early recovery. We have already established that a child with eczema appearing in the first year of life has about a 50 per cent chance of recovery within four years; on the other hand, if the eczema first appears at the age of five, there is probably only a 25 per cent chance of the eczema recovering within a further four years, and for a 50 per cent

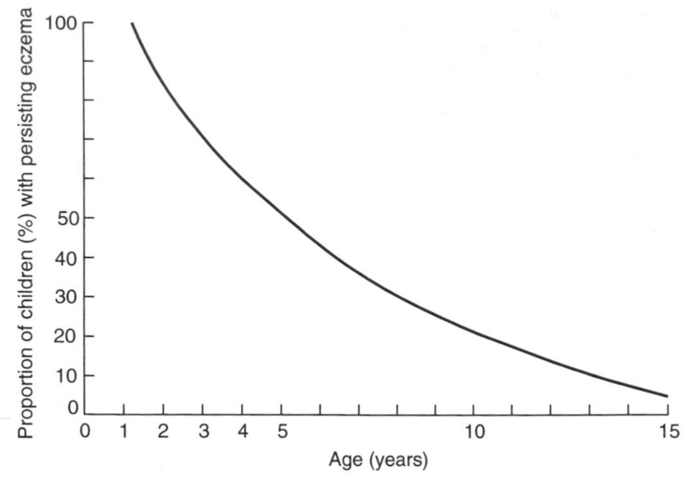

Figure 31 This graph shows the statistical chances of atopic eczema clearing up in children with an onset in the first year of life. You can see that there is about a 50 per cent chance of resolution by the age of five years.

chance of recovery the child may have to wait until the age of 15. There is no doubt that the outlook is adversely affected by a later onset, particularly an onset in adolescence or in adult life.

A problem for anyone who has recovered from atopic eczema is that they are always at risk, albeit a small one, of getting eczema again later on. If one talks to adults with atopic eczema, it is surprising how often the story goes as follows. The eczema was present but trivial in early childhood, and subsequently more or less disappeared. But later, around the age of ten, it reappeared in a more severe form and then persisted. In other adults, recurrence of eczema can be provoked by an unsuitable job or leisure activity; this is considered further in Chapter 12.

Having indulged in some generalizations, it is important to remember that the course of any person's eczema is peculiar to that individual, and is largely unpredictable. Any individual may do better or worse than the average. I have given some idea of the overall statistics, such as they are, and they are in many ways fairly comforting. Indeed, a child's treatment must at all times take into account the great likelihood of eventual recovery.

Parents often worry what their child's skin will be like when the eczema goes. They are usually concerned about possible scarring, and, when their child's eczema is severe, this is very understandable. It is, however, one of the great consolations of eczema that it only extremely rarely leaves scars. This is because the damage occurs quite close to the surface of the skin, however severe it is. This is something that one therefore needs to worry about very little.

Another type of scarring which parents naturally fear is 'psychological scarring'. They feel that years of eczema cannot fail to take a toll on their child's mental health. Fortunately, this also turns out to be something about which they need not be too concerned. When the eczema clears, most children will quickly forget that they ever had it, even in cases where it has been severe, and they will be psychologically as normal as any other child. Where children do suffer behavioural difficulties later on, these do not generally arise as a direct result of the eczema but indirectly because of the way they have been handled by parents and by others. This is why, as far as possible, eczematous children should be treated exactly the same as any other child, why they must never be regarded as 'delicate' or 'different', and why they should *never* be spoiled in compensation for their disease. This is an important subject, to which we will be returning in Chapter 11.

I urge all parents to remember that, however desperate things may seem, in the vast majority of children eczema eventually recovers, and does so without any permanent harm being done either to the skin or to the psyche. This is very different from most of the other chronic diseases that bring children to our hospitals, and is something from which to take continuous encouragement.

7

Skin infections

Skin infections are common, unpleasant, and occasionally dangerous complications of atopic eczema, which can cause considerable difficulties in diagnosis and treatment.

Broken, damaged skin provides an attractive environment in which many kinds of micro-organism can flourish and multiply. These include bacteria, fungi, and viruses, each of which I will consider in turn.

BACTERIAL INFECTIONS

Bacteria are micro-organisms consisting of single cells which are somewhat different from other animal or plant cells. They depend on their environment for nourishment, and sustain themselves in a wide variety of ways. In many respects, bacteria have been the most successful of all living creatures because they have been able to adapt to the greatest variety of habitats. Bacteria have even been found which can survive in the baths used to strip paint from pine furniture! However, each type of bacterium has a favourite habitat and may not be able to survive elsewhere. When conditions are favourable they multiply by dividing into two. Because they can divide so quickly, their population can increase at an alarming rate.

Certain bacteria can be found on the skin surface in entirely normal people and they do no harm. On the contrary, these bacteria, known as the *normal flora* of the skin, are almost certainly beneficial. Most bacteria make antibiotics, which they secrete into their surroundings to reduce the competition. In this way, the normal flora probably give us a degree of protection against harmful bacteria and fungi.

Some otherwise healthy people also provide a home on their skin for certain bacteria which are potentially damaging, but which cause no problems as long as the skin is not broken. Such persons are considered to be *colonized*, rather than *infected*, by these bacteria. The term *infection* is reserved for situations in which bacteria are actually doing harm to their host. Colonizing bacteria of this type, and those of the normal flora, tend to be present in very small numbers because the healthy, intact skin surface is an inhospitable place. It is dry and covered with inedible waxy scales. For this reason, bacteria living on healthy people tend to favour moister sites as are provided by the inside of the nose, the genital region, and the area between the buttocks. Any break in the skin covering will provide the moisture and nutrients they need in order to proliferate. Multiplying bacteria are a potential threat because they produce a variety of poisonous chemicals known as *toxins*. Some of these toxins help them to pass through the surface into the body itself, with potentially disastrous consequences

The function of the immunological system is to back up the protection provided by the skin and internal body surfaces, in order to prevent harmful bacteria from breaking through and invading the body proper. Wherever such bacteria start to break through, the immunological system will fight back to eliminate them.

Eczematous skin provides ideal conditions for many types of bacteria to flourish, and a difficult battleground for the immunological system. The result is, at best, a new balance between the bacteria and the immunological system, in which larger numbers of bacteria are present while the situation is still kept under some sort of control. In practice, the balance tends to be a rather uncertain one, with the numbers of bacteria sometimes being controlled and at other times not. While the situation remains under control, doctors would still see it as colonization, but when the population of bacteria increases to the point at which the skin is actually damaged by their activities, this would be regarded as infection.

Staphylococcus aureus (*Figure 32*) is one of thousands of types of bacterium which can cause infections in humans, but it is a specialist when it comes to skin infections. It is responsible for most cases of impetigo, a common skin infection which can occur in otherwise healthy children. It is also the usual cause of boils and styes. *Staphylococcus aureus* is responsible for the great majority of bacterial infections of the skin in anyone with eczema. It now seems clear that these bacteria increase the damage in eczematous skin when their population exceeds one million per square centimetre. Almost all people who have eczema will have *Staphylococcus aureus* on the skin somewhere. Unless the population is high, they probably do little harm. However, infections with *Staphylococcus aureus* (*staphylococcal infections*) appear to be the commonest cause of sudden worsening of eczema. The problem is recognizing when the skin is infected. Infection requires treatment because the bacteria involved are actively damaging the skin, whereas colonization alone does not require treatment as the bacteria are doing no harm. In practice, the problem is to tell the difference.

Infection with *Staphylococcus aureus* should be suspected wherever there is worsening of eczema associated with weeping and the

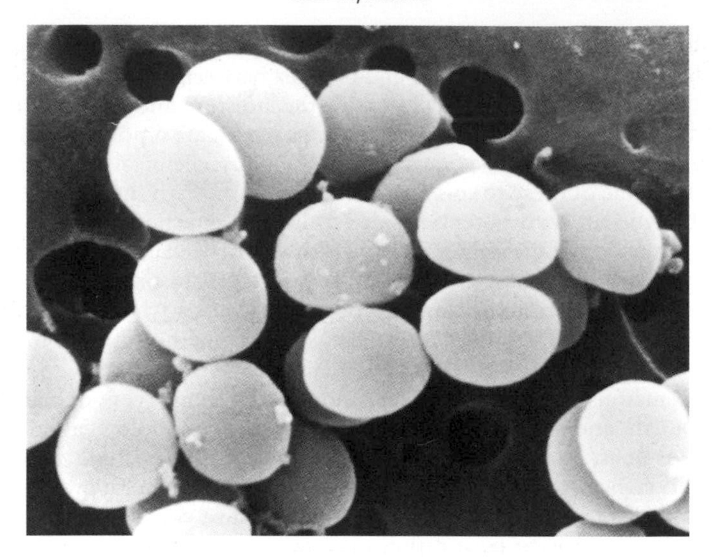

Figure 32 *Staphylococcus aureus* bacteria, the most frequent cause of skin infections in children with atopic eczema, magnified 20 000 times (scanning electron micrograph reproduced by kind permission of SmithKline Beecham Pharmaceuticals).

appearance either of crusts with a particular yellow hue (*aureus* is Latin for *golden*), or pus-filled spots (*Figure 33*). These changes in the skin are often combined with painful enlargement of lymph nodes and, occasionally, with fever. *Lymph nodes* are the lumps, sometimes painful, that parents will often notice in their children's groins, axillae (armpits), and necks. Their job is to filter out bacteria, yeasts, and viruses from them to prevent them entering the blood. When they are having to work really hard they often become enlarged and tender (*Figure 34*), and although parents are often worried by this, it should really be seen as a sign of health. They are a witness to a vigorous effort to protect your child from infection of the blood.

Unfortunately, these obvious visual signs of infection of eczematous skin by *Staphylococcus aureus* are not usually apparent until the population density of the bacteria reaches about ten million per square centimetre, that is, ten times the population at which aggravation of eczema occurs. This means that infections that worsen the eczema may not be detectable at all by eye. Ideally, special tests would be available to detect the presence of infection when it is still undetectable visually.

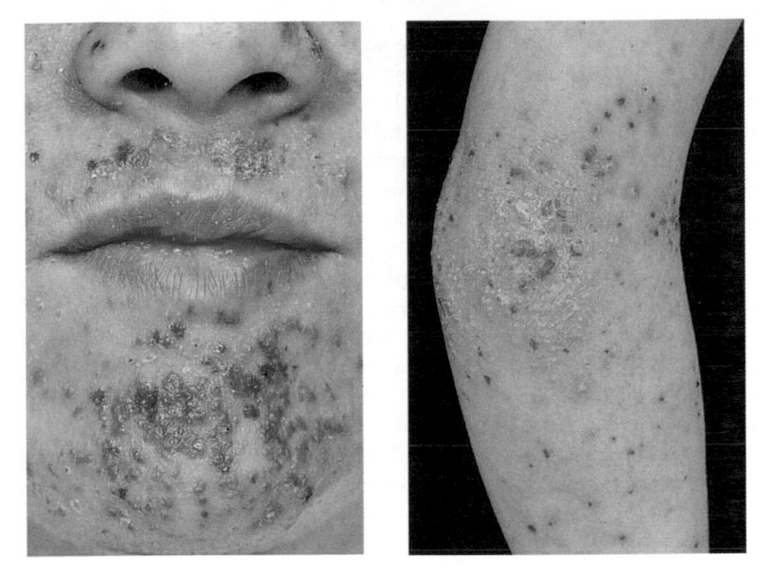

Figure 33 Atopic eczema with features of bacterial infection.

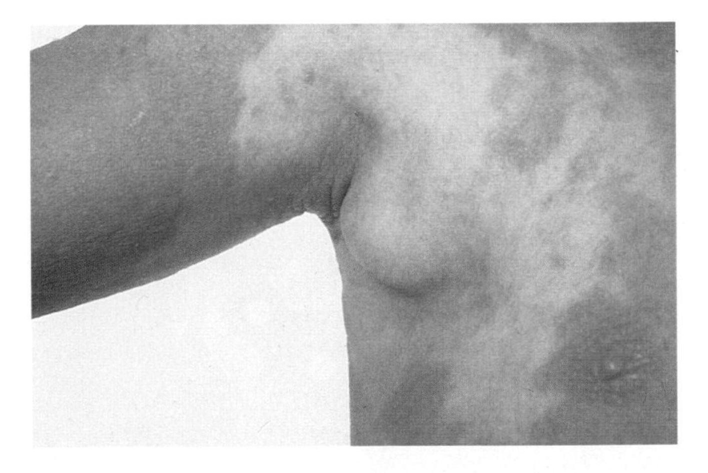

Figure 34 An obviously enlarged lymph node under the arm.

Doctors often take *swabs* from the skin if they suspect an infection. As a general rule, this involves rubbing a moistened cotton-wool ball into the affected area, then sending it to a hospital laboratory. In the laboratory, this material is in its turn rubbed on to a special plate of nutrient jelly known as *culture medium*, and the plate is placed overnight in a warm *incubator*. The next morning the bacteria will have formed little spots on the jelly, known as *colonies*, just like mould growing on food (*Figure 35*). The type of bacterium can be recognized by the shape, texture, and colour of these colonies, and by studying the bacteria within them under a microscope. The problem with this technique is that it does not allow an accurate estimate to be made of the population density of the bacteria on the skin. When the population density on the skin is very high, the number of bacterial colonies growing on the culture plate is likely to be high, but this is not always the case and in practice the method does not allow accurate quantification of the bacterial population on the skin. There are ways of doing more accurate counts of bacteria on the skin, but these techniques are not widely available. Currently, therefore, examination by swabbing and culture only tells us if bacteria are present, and which bacteria they are, but not their number. The technique is therefore relatively uninformative.

Some types of bacteria should not be present in the skin at all, even in those with skin disorders, and their discovery by swabbing and

Figure 35 A laboratory culture plate growing colonies of *Staphylococcus aureus* bacteria (reproduced by kind permission of Professor W. Noble, St John's Institute of Dermatology).

culture requires that the patient be treated with the appropriate antibiotic. A good example of this type of bacterium would be *Group A streptococcus*. This bacterium can occasionally cause the potentially serious heart condition known as *rheumatic fever* or an equally serious kidney disorder called *acute nephritis*, so that a child with these bacteria on the skin is personally at risk and is a threat to others. I should perhaps add by way of reassurance that these complications are nowadays extremely rare. Its detection by swabbing and culture almost invariably indicates that an infection is present, so that quantification is not necessary. The commonest type of problem caused by group A streptococci is however not skin infection, but sore throats, and if your family are getting sore throats, it can occasionally reflect skin infection by this bacterium in your child with eczema.

Thus, taking ordinary swabs can be helpful because it can detect certain particularly unpleasant bacteria, but it is of very limited value in the case of infections or possible infections with the much commoner *Staphylococcus aureus*. However, it does have one use in relation to this and other bacteria, by determining the sensitivity of the bacteria to individual antibiotics. Antibiotics can be put on to the culture plates, and their effect on the growth of colonies can be measured (*Figure 36*).

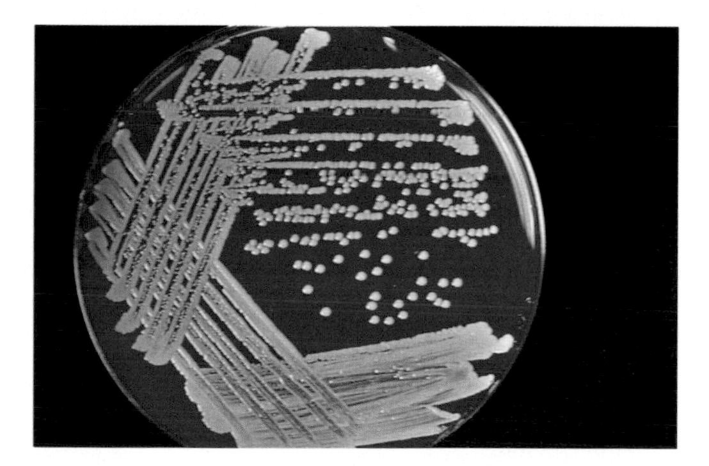

Figure 36 Paper discs containing various antibiotics have been placed on this culture plate to assess the sensitivity of *Staphylococcus aureus* bacteria to their effects. With this method, one can check that an infection will respond to the antibiotic that the patient has been prescribed (reproduced by kind permission of Professor W. Noble, St John's Institute of Dermatology).

This helps doctors know which antibiotics will be most appropriate for a particular infection.

I want to return to the difficulty of knowing whether a person with eczema has a skin infection due to *Staphylococcus aureus*. I have explained that infection may be present without it being visually obvious, and that the swab and culture technique is of very limited help in telling one whether infection is present rather than just colonization. Until we have better tests to assist us, we will remain unable to make this distinction accurately. In practice it is wise to assume that worsening of eczema which occurs over a fairly short space of time, and which is not readily explained by other factors, is likely to be due to such an infection. Because this will not always be the correct assumption, patients will undoubtedly be liable to excessively frequent treatment for infection.

Treatment of bacterial infections is discussed in Chapter 8. However, in general terms, treatment is largely with antibiotics, which are more effective when given by mouth or injection rather than by direct application to the skin. Most people who have atopic eczema will experience bacterial skin infections from time to time. Many will at some time suffer frequent and repeated infections, which usually result in multiple courses of antibiotics being prescribed one after the other, often without preventing a recurrence of infection a few days later. This does not generally happen because the bacteria are passed on to the eczema sufferer from other people but because the conditions in the skin are so encouraging to the small numbers of bacteria that manage to survive the antibiotic that they rapidly reproduce, until the population is again large enough to cause infection. On other occasions, the bacteria will reinfect the eczema sufferer from someone close to the sufferer who is colonized, a very frequent state of affairs in the immediate family of anyone with eczema. The usual site for such colonization in those with healthy skin is the inside of the nose, from where the bacteria can easily get around on the fingers. If frequent infections are a problem, it can be worthwhile to seek treatment for this type of colonization in family members or close friends. Sometimes, family members will themselves get full-blown skin infections, usually in the form of impetigo, crops of pus-filled spots, boils, or infections in cuts, abrasions, and so on.

The most important treatment for skin infections in eczema is prevention. Antiseptics have very little place in this respect, because they are generally irritating to the skin unless they are used in such dilution that they are ineffective. Effective prevention involves making the skin as unattractive as possible from the point of view of bacteria. This means good skin care, particularly frequent baths or showers, using appropriate oils in the water, and creams to clean the skin physically in place of soap. It also means the frequent application of moisturizers at other times, to seal the surface of the skin with a film of oil, which is very much to the distaste of bacteria.

FUNGAL INFECTIONS

Fungi are almost as widespread in nature as bacteria. By and large they cause little problem to people, except when they attack our food. Fungi were responsible, for example, for the potato famine that caused millions to die in Ireland in the 1850s, and were responsible in the last decade for the disappearance of elm trees in much of the UK.

There are two principal types of *fungus*. *Yeasts* are the first type; they resemble bacteria in comprising single cells, but these cells are larger and more sophisticated. They divide in much the same way as bacteria in order to reproduce. Yeasts are a special variety of fungus that can also cause problems, but which has also served us well by playing a crucial role in making wine, beer, bread, and many cheeses.

The second type of fungus comprises chains of connected cells known as *hyphae*, which form thread-like tubes that they use to invade their habitat; these fungi are called *moulds*.

Certain yeasts live on the skin of perfectly healthy people as part of the normal flora, in the same way as some kinds of bacteria. Like them, they probably do good by producing antibiotic substances that deter other bacteria and fungi which might be harmful. Of these, the best known are the *Pityrosporum* yeasts, which we have already considered at some length in Chapter 5. As far as we are aware, they do not tend particularly to cause infections in those with atopic eczema. However, they are now believed to be able to aggravate atopic eczema in some cases, when an individual develops IgE antibodies which can then provoke an allergic reaction against them (p. 57).

Certain types of fungus of the mould type are a frequent cause of skin infections in the general population. These fungi are called *dermatophytes*, and they are responsible for common infections of the scalp and hair (*scalp ringworm* or *tinea capitis*), and of the body (*ringworm* or *tinea corporis*) in children. These fungi do not cause infections which are either more frequent or more troublesome in children with atopic eczema.

VIRAL INFECTIONS

Viruses are another type of micro-organism, and are very different from bacteria and fungi. They are very much smaller, and do not comprise whole cells. They consist of little more than an envelope containing some DNA or a similar protein called RNA. The DNA or RNA are able to make copies of themselves only if they can get to the necessary equipment inside the cells of their chosen host, the equipment the host normally uses to reproduce its own DNA and RNA. Once inside the cell, this equipment may more or less be taken over by the virus to make vast numbers of new virus copies. Viruses use all sorts of clever techniques to get from one cell to the next.

Certain viral infections occur in more or less everyone at some time during childhood, particularly *herpes simplex, warts, mollusca*, and *chickenpox*. This also used to be the case with *measles, rubella* ('German measles'), and *mumps*, but these are now becoming extremely rare in developed countries because of changes in the immunization programme. In general, once one has had any of these, one develops immunity which protects from second attacks. However, the level of protection varies somewhat from infection to infection. Second attacks of warts, for example, are not uncommon, while a second attack of mollusca is extremely rare.

ECZEMA HERPETICUM

Certain viruses are able to take advantage of the absence of the normal skin defences in a person with eczema, and can establish an infection from which they can easily be spread from cell to cell during their victim's bouts of scratching. These infections may sometimes be dramatic, and are occasionally dangerous. The virus that most commonly does this is the *herpes simplex virus* (*Figure 37*). The infection it causes in those with atopic eczema is called *eczema herpeticum*, and today, this is more or less the only serious viral infection of eczema.

Herpes simplex virus (or HSV, for short) is the cause of the common *cold sore* that many people get around their lips from time to time. Very few escape an infection with this virus at some time during childhood or early adult life. However, once one has had such an infection, an immunity is developed which provides protection from further attacks.

In those who do not have eczema, the initial infection with HSV is most commonly in the mouth and on the lips, where it causes crops of vesicles and ulcers, which may be very painful. Most of these initial infections are never recognized for what they are, because they are usually mild, and may not be noticed at all. They may be more severe and simply passed off as 'mouth ulcers'. During this first infection, the immune system works fast to get the better of the virus. With an

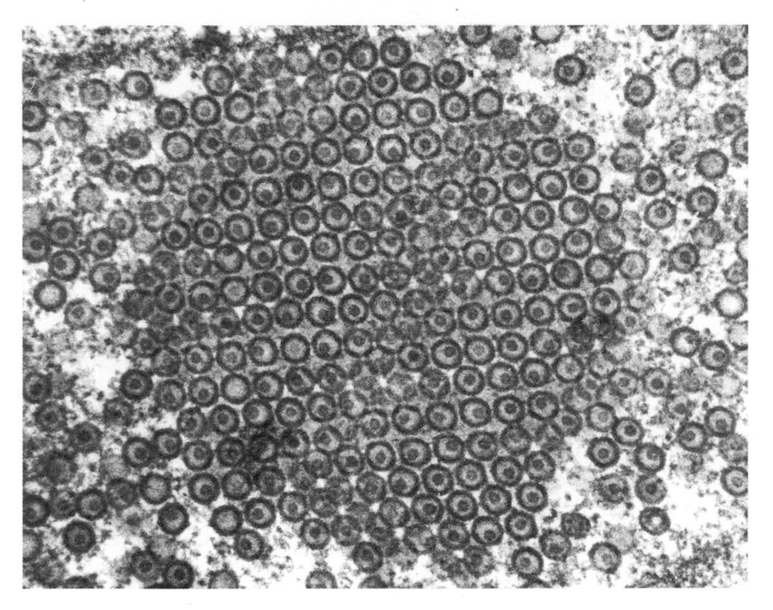

Figure 37 Herpes simplex viruses within a cell, magnified 30 000 times (electron micrograph reproduced by kind permission of SmithKline Beecham Pharmaceuticals).

infection that it has not previously encountered, this takes a little longer, and how long it takes is probably the main factor determining the severity of the infection. However, once the initial infection has been overcome, the immune system is able to memorize essential details of the virus and will be able to deal with future encounters much more swiftly and effectively.

Because of this memory function of the immune system, HSV reaching the body from the outside can be mopped up quickly on future occasions, and will fail to become established. However, small numbers of HSV often linger on from the initial infection, hidden away from the body's defences. Their favourite hiding place appears to be the nerves that supply sensation to the area where the initial infection occurred. Since this is usually in or on the lips, the nerves infiltrated by these 'guerrilla' viruses are those in the skin around the lips. Most of the time the immune system keeps them locked away in these sites, unable to venture further. However, when the body is run down by another illness, for example, these viruses take the opportunity to travel down the nerves into the skin around the lips, where they set up a localized infection. When the immune system realizes what is happening the infection is quickly dealt with, but not before the victim has had an uncomfortable cold sore.

Unfortunately, small numbers of viruses may hold out for years in their outposts in these nerves, so that infections of this type may be a recurrent problem. They are known as *herpes labialis*, or cold sores (*Figure 38*). Reactivation is mainly provoked by strong sunlight on the lips, or even just a cold wind, which seems to be enough to weaken immune defences in the area.

The problem with cold sores is that they provide a source of viruses that can start off infections in children who have never encountered them previously, and who have therefore no pre-existing protection. Cold sores are the source of virus for new infections in susceptible individuals, usually young children, to whom they are most often transferred by the contact of kissing.

Although the initial infection with HSV is normally in the mouth, in children with certain skin diseases, especially eczema, it may take place instead on affected areas of skin. If the eczema is severe and extensive, the infection can spread rapidly, and the situation can become a serious one. Infection of eczematous skin with HSV is known as *eczema herpeticum*.

If this happens to your child, you may not immediately realize the

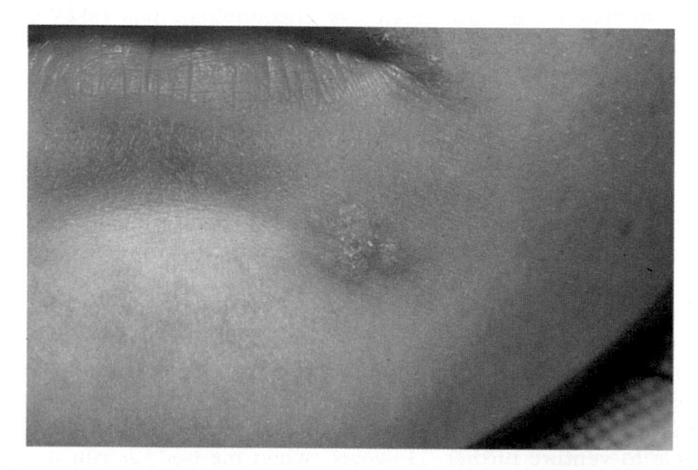

Figure 38 A 'cold sore', the commonest source of herpes simplex virus.

(*a*)

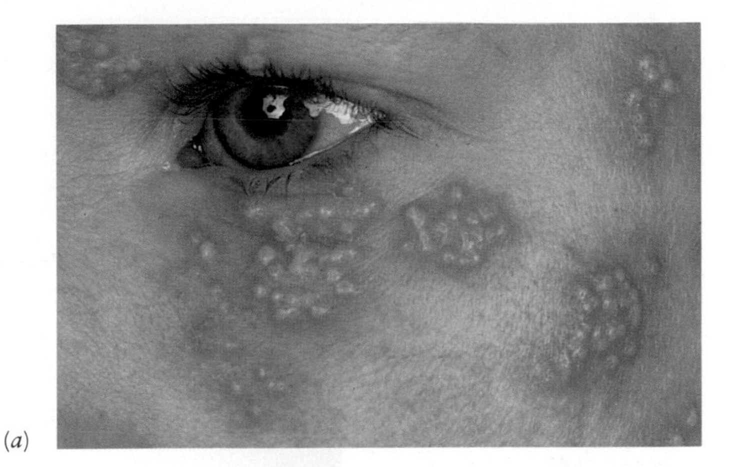

(*b*)

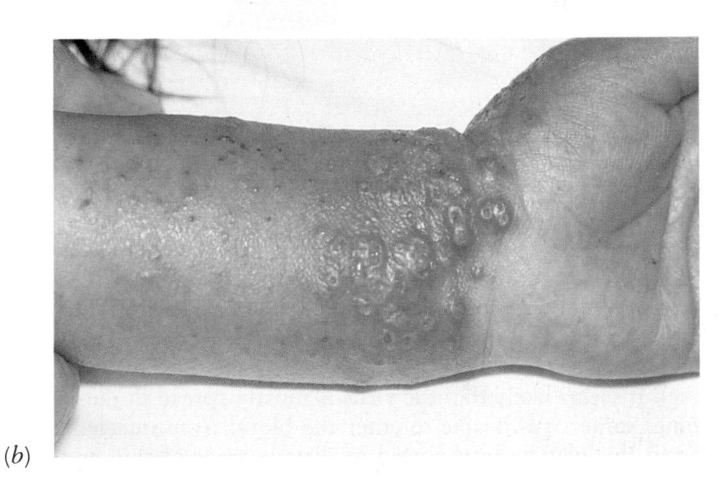

Figure 39 Eczema herpeticum (*a*) Typical clusters of umbilicated blisters on the face at the earliest stage of the infection, (*b*) a less extensive outbreak on the wrist.

cause of the problem. The virus infection causes a rash which is superimposed upon the eczema itself. This rash consists of clusters of vesicles, each about 2 mm or 3 mm across (*Figure 39*). At first they are filled with clear fluid; later this often turns to pus. Many of the blisters develop a small central depression, which is highly characteristic. The

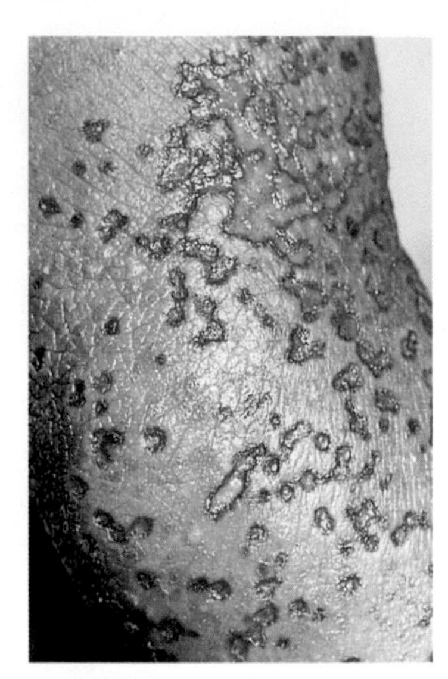

Figure 40 Eczema herpeticum: the characteristic punched out ulcers that follow the earlier small blister stage. Note that each of the lesions is of a very similar size.

tops of these rather fragile blisters tend to be scratched away fairly quickly, leaving raw, weeping, or pus-filled areas, which later become crusted over (*Figure 40*). The most common place for this infection to start is on the face, where lesions have a special tendency to congregate around the eyes, and on the hands and fingers. From the time one sees the first vesicles, the infection tends to spread and intensify for about four or five days, after which the immune system generally gets to grip with it. At this stage the new vesicles cease to appear, and gradually the whole infection resolves.

Though it seems likely that the virus is mostly spread in the skin by scratching, some virus is able to enter the blood from infected areas, and can in this way be transported to distant areas of skin and also, occasionally, to internal organs. This is what makes eczema herpeticum potentially dangerous. Without wishing to cause unnecessary alarm, I feel that readers should know that the condition is occasionally lethal. Fortunately, this is extremely rare. Most attacks of eczema herpeticum are mild, and limited in extent. In fact, it seems likely that most of these infections are so mild that they are not actually recognized.

As with infections in those without eczema, recurrences may occur in anyone who has had eczema herpeticum. It is similarly very unusual for these to be as bad as the initial infection (*Figure 41*).

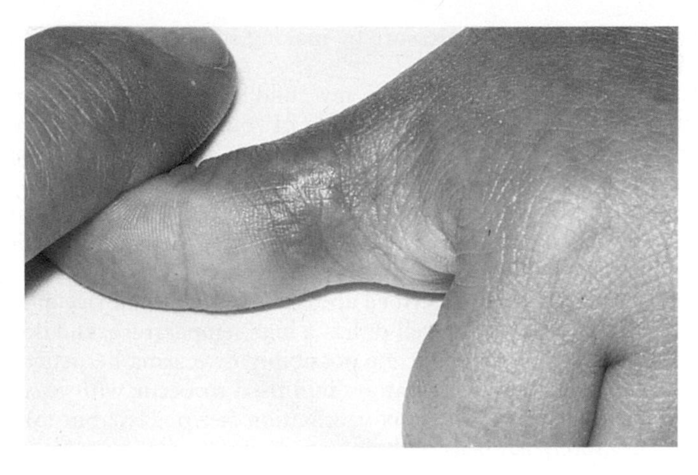

Figure 41 Eczema herpeticum: typical localized recurrence at a previously affected site.

A drug is now available, known as *acyclovir* (trade name Zovirax®), which can control the multiplication of this virus. It can be given by mouth, but works more rapidly and effectively when given intravenously by drip in hospital. If it is given early enough, no child should ever die from this infection.

However, treatment with this drug is not always necessary. In particular, it is not necessary to treat a child after the fourth or fifth day, if new vesicles are no longer appearing, because by this time the body itself is in control and, in any case, the drug only works while the virus is rapidly reproducing itself early in the infection. There is a good case for giving the drug early in the infection, even if it is not extensive, as one cannot know at this stage just how bad it will get. One can give the drug by mouth if the child will take it, and if one feels fairly happy that it is not likely to become severe. If there is any cause for anxiety, the acyclovir should be given intravenously, which will mean a brief stay in hospital.

Acyclovir applied to the skin may have a slight effect in eczema herpeticum, but because the spread of infection is partly through the blood, the drug is best given internally if it is used at all. However, for recurrences, local application of acyclovir can be very helpful if used quickly enough, before the virus has really got going. It is also important to keep up the pressure by making sure that it is applied at least five times daily.

Clearly, you should try to keep any child with eczema away from anyone who has a cold sore. If a member of your family ever does have one, it should be kept covered and that person should avoid physical contact with the eczematous child until the cold sore has gone. If your child's eczema ever rapidly worsens for no apparent reason, the reason may be HSV infection. Suspect it particularly if there is an obvious change in the quality of the eczema, and especially if you see the clusters of small blisters I described above. Seek medical advice quickly if your child is obviously unwell or has a high temperature, and do not be afraid to ask directly about the possibility of eczema herpeticum.

A similar infection in eczematous skin used to occur with *vaccinia* virus, the virus used for smallpox vaccination (see p. 218), but fortunately it very rarely happens with any other virus.

CHICKENPOX (VARICELLA)

Chickenpox is spread in nasal or oral secretions. After an incubation period of about two weeks, the child may feel feverish and unwell, though many do not. The characteristic rash of small blisters (known as *vesicles*) makes its appearance. These appear in crops, initially in skin that looks normal, but quickly there develops a surrounding area of redness. Most of the vesicles occur on the trunk, with smaller numbers on the limbs, particularly those parts nearest the trunk. Some lesions occur in the mouth. The number of lesions is very variable, but tends to be higher in children with eczema than in other children.

The lesions can be quite itchy, and though it is frequently asserted that scratching the tops off them leads to scarring, I think that it probably makes very little difference whether they are scratched or not. Sadly, chickenpox often causes more permanent scars than many years of eczema.

Oddly, children with eczema often experience improvement of their eczema while they have chickenpox, often more or less clearing up altogether, but usually for no longer than a week or two. The same phenomenon was often noticed when measles and rubella were still common. The reason seems to be that the cells known as T-lymphocytes, which appear to cause much of the trouble in the skin in eczema, are also the cells responsible for getting rid of viruses. During these infections, the T-lymphocytes appear to become so busy dealing with the viruses that they are unable to keep up their less helpful and self-destructive, eczema-generating role.

WARTS

Warts occur in almost all children at some time. In most cases, the infection probably goes unnoticed, as there are only one or two small lesions, often hidden away on the soles of the feet (people often call warts on the soles *verrucas*, though this is in fact the Latin word for *all* types of wart). They are transmitted from child to child by direct contact, or through some third-party object, such as the floor in the case of warts on the soles. Warts last for no longer than a few months in most cases, and then disappear, as a result of the development of immunity against them. This immunity provides protection from further attacks in most cases, but a few people do not seem to be able to develop long-lasting immunity against the wart virus, and continue to get warts that last for many years, or come back soon after they seem to have resolved. Very often, this poor immunity seems to run in families.

Children with atopic eczema can get warts like everyone else. However, quite often they get much larger numbers of them (*Figure 41*). This seems to be a result of the fact that they spread them about by scratching, rather than reflecting any failure in their immunity. Generally, they get rid of the infection just as quickly and efficiently as other children, and are no more likely to get another infection in the future.

As a rule, warts do not require any treatment. The treatments that are most likely to work are destructive and painful, and, even then, not that reliable. I would advise the parents of any child with eczema that has warts to avoid treatment, unless the warts are painful enough to interfere with normal activities, a situation that is more or less restricted to warts on the soles, or if they are causing substantial cosmetic handicap. Sadly, the more warts there are, the more likely a child or parent will want treatment, but the more difficult it will be to have any real chance of treating them successfully in this type of situation. Treatment will mean many painful sessions, which most pre-adolescent children will not, and should not, be asked to tolerate.

MOLLUSCA

Mollusca are in many ways similar to warts. Their full name is *Mollusca contagiosa*. The virus that causes them is transmitted from child to child, but almost always directly by touch. The lesions it causes are a little different, consisting of almost spherical spots that sit on the skin, looking just like pearls (*Figure 42*). As they mature, they often develop a central depression, out of which a core is eventually expelled. As they reach this final stage, they become itchy and are often accompanied by

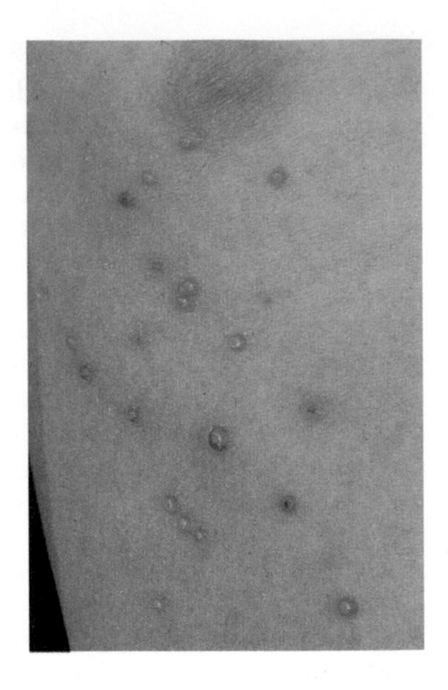

Figure 42 Mollusca contagiosa in a child with eczema.

inflammation in the surrounding skin which itself closely resembles eczema. After this, they generally disappear, and recurrent infections in adults are virtually unheard of. The whole process usually lasts a few months.

Children with atopic eczema tend to get more extensive infections, just as in the case of warts, because they spread the infection about by scratching. As with warts, the mollusca will go away in a few weeks or months, but in fact they will do so more reliably, and it is rare for infections not to have disappeared within a year, and almost unheard of for them ever to recur. Treatment is not generally advised, as it is almost always painful and often ineffective.

RISKS OF TRANSMISSION OF INFECTION

In general, it is possible for all the micro-organisms that cause skin infections in eczematous children to be transmitted to and sometimes to cause infections in others. As far as the virus infections are concerned, these are all common, and other children will get their infections from someone else's child if they don't get them from yours. The situation is rather different in the case of the bacteria that most commonly cause skin infections in eczematous skin. These bacteria, particularly *Staphylococcus aureus* and group A streptococci, can cause infections in those who do not have eczema but have close contact with an eczematous child.

Staphylococcus aureus causes the infection commonly called impetigo, which usually takes the form of yellow, crusted spots which can occur at almost any site, but most often on the face around the nostrils and mouth. This bacterium can also cause large and small boils, or infections in cuts and abrasions. While family members are more liable than others to get such infections from a child with eczema, friends and others who have close contact with the child, such as childminders, nannies, and babysitters may also pick up infections of this type. Such infections are rarely much of a problem, but recognition of the risk should accelerate their diagnosis and the start of appropriate treatment.

Staphylococcus aureus can also cause food poisoning if food becomes contaminated by it. There is a definite risk that this can happen when someone with hand eczema prepares food, unless they are aware of the risk and extremely careful. This is one genuine reason for anxiety about those with hand eczema in the catering industry.

Group A streptococci can be more troublesome. These bacteria can cause sore throats and tonsillitis, which are quite a common hazard of caring for eczematous children. If you develop a nasty sore throat, without the nasal symptoms that usually go with a cold, but associated with a temperature and feeling generally unwell, these bacteria may be responsible, and rapid cure could follow a brief course of penicillin from your GP.

Another issue is the infection risk that others may pose to children with eczema. As we have seen, eczematous skin is devoid of much of its normal protection against the entry of infectious micro-organisms. Children with eczema will therefore pick up skin infections of all types from other children with particular ease. But another important possibility, which tends to be discussed rather rarely, is that a person with eczema will be at risk of picking up types of infection other than ones that primarily affect the skin because the responsible micro-organisms are able to gain access too readily through the eczematous skin. Worries have been expressed that the hepatitis B virus and even HIV (the cause of AIDS) might be transmitted this way, whereas in those with normal skin they would generally have to be introduced by injection. There is little for the usual eczema sufferer to fear from this because they are so unlikely to have adequate skin contact with the body fluids of a person carrying such viruses. However, the issue does become more worrying when someone with eczema is nursing someone with one of these infections, and may get blood or vomit on their hands for example. This is another reason why nursing may not be a very good idea for those with hand eczema.

8

How should eczema be treated?

My aim in this chapter is to describe the spectrum of treatments available for atopic eczema in children. In practice there is quite a wide range of treatments which offer considerable choice. However, treatments that suit one child may not suit another. The 'right' treatment for the individual will not be one treatment but a package of treatments which provide an appropriate balance for that child.

Each eczematous child is different, both in the quality and quantity of the problem, and treatment must be tailored to fit the needs of that individual. Eczema must be neither undertreated nor overtreated. It can be difficult to be sure initially what will suit a particular child, and the ideal treatment combination is often arrived at only by a long process of trial and error. Even then, what is 'ideal' will keep on changing over the months. Complete eradication of the eczema is not a realistic therapeutic objective. What one hopes to achieve is an adequate level of control.

To achieve this level of control requires constructive collaboration between parents, child, and doctor. It is unreasonable to place full responsibility for treatment on your child's doctors, and you may lose their support by demanding miracles which cannot possibly be performed. Parents share responsibility for treatment, and its success depends largely on their efforts. They must take much of the credit when treatment is successful, and must ask themselves if they do not deserve at least some of the blame when it is not.

I see the function of this small book mainly as one of giving information. I want parents to share what is known about eczema and how to cope with it. It is worth reflecting that if we really knew all there was to know, there would be no need for this book at all.

Please note the symbols I have used in this chapter to denote the availability of different treatment products I have mentioned in the text:

’ only available on prescription, either from your GP or your hospital specialist
* can be prescribed or bought
+ not available on prescription, but can be bought or can be supplied by hospitals

^ *only* available if you buy it yourself
# can only be obtained on the NHS by a hospital specialist's prescription, or from your GP on a private prescription for which you would have to pay

TOPICAL APPLICATIONS

CREAMS, OINTMENTS, OILS, LOTIONS, AND GELS

These are some of the terms used for treatments designed to be applied directly to the skin. They are correctly known as *topical, topical applications*, or *topical treatments*. These names refer to topical applications having different physical properties that make each of them more suitable for certain jobs than for others.

Ointments and *creams* are different from one another, though they are often confused, even by doctors, nurses, and pharmacists. Ointments generally comprise oil alone, almost always mineral oil of a grease-like consistency, usually predominantly white soft paraffin (like Vaseline®), though other oils may be added to give a particular consistency or 'feel'.

Creams, on the other hand, are principally water, to which oil has been added. As oil and water will not ordinarily mix, *emulsifiers* must be added, to make the oil break up into tiny droplets which remain stable and do not tend to coalesce. Because the result is a *suspension*, or an *emulsion*, of oil in water, the water is said to provide the *continuous phase*.

Lotions are just very watery creams, with a higher proportion of water to oil. Water is the continuous phase.

There are several variations on these themes. The term *oily cream* is used to describe ointment to which water has been added, and emulsified so that we now have a suspension or emulsion of water droplets in oil. In this case, the continuous phase is oil.

Some applications do not fit into these categories. A good example is a *gel*. Gels consist of water in which some other substance forms a three-dimensional lattice. The type of jelly we eat is a good example of a gel, in which the lattice is provided by a protein called collagen. Of course, for skin treatment the gel has to be much sloppier. Perhaps the best-known example of a gel which is useful in the treatment of eczema is *Unguentum Merck®.

Bath oils are, as the name implies, oils designed to be used in the bath. They are generally also suitable for the shower. They are of two general types, *spreading* or *dispersing*. The spreading types float on the surface of the water, and a layer is therefore picked up as one gets in and out of the bath, whereas the dispersing types have added emulsifiers, so that they break up into millions of tiny droplets as they mix with the water. One can usually tell the difference quite easily as the dispersing types become white as they mix with water. A good example of the spreading type is *Balmandol®. Examples of the dispersing type are *Balneum®, *Alpha Keri®, *Oilatum emollient®, *Emulsiderm®, and *Hydromol®. In fact, many of the dispersing types have a spreading component as well, and this is one of the reasons why they all function differently in practice.

ADDITIVES IN TOPICAL APPLICATIONS

Preservatives

Where there is a risk that any of these preparations could support the growth of micro-organisms, particularly bacteria and moulds, something has to be added that will prevent this or at least slow it down. This is the function of *preservatives*, of which there are several different kinds. In general, bacteria and fungi dislike mineral oil as they cannot use it as a food source. There is therefore generally no need for most ointments and bath oils to contain preservatives. Creams, gels, and lotions will however require added preservatives. In general, the greater the water content, the greater will be the need for preservative. Thus lotions will need more preservative than creams and gels, and little or no preservative will be needed in oily creams.

The problem with preservatives is that they can themselves sometimes cause aggravation of eczema. They may do this by provoking allergic reactions (p. 94), or by a direct irritant effect. In children, genuine allergic reactions to preservatives are rare, and irritant effects a much commoner explanation of the aggravation of eczema they may occasionally cause.

Commonly used preservatives in topical applications include parabens, chlorocresol, sorbic acid, and phenoxyethanol. All of these may occasionally cause allergy, but do so very rarely in relation to their very widespread use in cosmetics as well as in medical topical applications.

Antioxidants

Many types of oil have a tendency to *oxidize*. This is the cause of *rancidity* in oily foods such as butter. *Antioxidants* are added to topical applications to prevent this. Commonly used antioxidants include vitamin E, butylhydroxytoluene (BHT), and gallate.

Emulsifiers

Emulsifiers are used to stabilize the oil-in-water emulsions in creams and lotions, and the water-in-oil emulsions in oily creams. Emulsifiers commonly used in topical applications include sodium stearate, calcium oleate, sodium lauryl sulphate, polysorbate, cetyl alcohol, sorbitol, and polyethylene glycol.

Chemicals known as *stabilizers* are often used in addition, to help maintain the stability of emulsions; a commonly used example is cetostearyl alcohol.

Lanolin

Lanolin (or *wool fat*) is a common component of topical applications and of cosmetics. It is obtained from sheep's wool. It is used as an emulsifying agent, and results in a pleasant 'feel' to a cream. Natural lanolin is a mixture of substances, such as wool alcohols, which may sometimes be used as additives in their own right. The problem with lanolin is that it may cause allergy. As in the case of preservatives, this is in fact rather rare in childhood.

Fragrance

Pleasant-smelling substances are very common components of cosmetics and are not infrequently used in topical applications. Sometimes, they are used to mask other smells, such as those that may originate in emulsifying agents.

ALLERGY TO ADDITIVES IN TOPICAL APPLICATIONS

Individuals may develop allergy to almost any of the additives used in cosmetics and medical topical applications. Lanolin and preservative allergies are perhaps the most common. The development of these allergies generally requires many years of frequent use on damaged skin. Those with atopic eczema seem to develop such allergies somewhat less easily than those with other types of skin problem, and they seem to develop particularly rarely in children.

The allergy caused by these additives may occasionally be of the contact urticaria type (p. 39), but more characteristically, it is of a type called *allergic contact allergy*, and the reaction it causes is called *allergic contact dermatitis*. The process of becoming allergic in this way is called *allergic contact sensitization*, or just *sensitization*.

Allergic contact allergy is different from what I earlier called 'atopic contact allergy' (p. 42). In general, it is a type of allergy that develops against chemicals, simple and complex (usually called *sensitizers*), which are not proteins, whereas atopic contact dermatitis is the characteristic type of contact dermatitis that develops against protein antigens. The end result of both types of allergy is dermatitis (i.e. eczema), they cannot always be easily distinguished. However, the causes and mechanisms are different in each case. Whereas tests for IgE antibodies, such as the skin prick test (p. 8), are useful in the case of atopic contact allergy, a different test, known as the *patch test*, is used to detect allergic contact allergy.

Allergic contact allergy is quite a common cause of skin problems throughout the world. Because allergic contact dermatitis may develop against many of the chemicals used in industry, it is a common form of industrial disease, a frequent cause of time off work in adults, with important social and economic consequences. It is the type of allergy that may develop against common non-protein-containing substances in our environment; well-known examples include nickel (in dress jewellery, metallic clothes-fastenings, and coins), rubber (in household rubber gloves, clothing elastic, and shoes), cobalt (in dress jewellery, metallic clothes-fastenings, and cement), and chromate (in cement, industrial cutting oils, and leather), but there are many others. Although allergic contact allergy is rare in children in Europe, it is not uncommon in the USA, where it is frequently provoked by two widespread plants, called poison ivy and poison oak. None of these types of allergy seems to be very much more common in those with atopic eczema, though the broken skin that is part of most skin diseases does predispose somewhat to sensitization because of the resulting failure of the skin's protective barrier function.

Whereas allergic contact allergy to topical applications is relatively rare in the general population, it is a very real problem in those with skin disease, both because of the defect in barrier function, and, more importantly, their frequent use of topical treatments containing the relevant sensitizers. However, as I have mentioned before, it is relatively rare in children, especially younger children.

Allergic contact allergy to cosmetics is also quite common in the general population, and is likely to be a special problem in those with long-standing skin disease, because the chemical additives that cause it are the same as those in the topical applications used for treatment.

If one suspects allergic contact allergy, the appropriate test is the *patch test*. In this test, the substance in question is applied to the surface of the skin, usually on the back, under a special dressing, and is left in place for 48 hours. If this type of allergy is present, a patch of eczema will appear at this site, which will still be there a further 48 hours after the test substance has been removed. The tests involve attendance at a dermatology out-patients' department on three half-days in a week, and are therefore a major undertaking for patients. Because this type of allergy is rare in children, such tests are not often undertaken during childhood unless there is a strong suspicion of allergic contact allergy. Patch testing should be contemplated in children with atopic eczema in whom there is strong circumstantial evidence of allergy to any of the known common causes of allergic contact allergy. It should particularly be considered when there is a major suspicion of allergy to topical applications. It is, however, important to remember that when atopic eczema appears to be aggravated by a topical application, the cause is very much more likely to be a direct irritant effect than a genuine allergy, and it can often be overcome by a change in the applications used. Children are often unable to keep the patches in place for the necessary length of time, and the tests cannot be undertaken in the first place if the back is affected by eczema. Furthermore, worsening of the eczema by the test patches may occur as a result of an irritant effect from the test substance, so that interpretation may be difficult. For all these reasons, such tests are only rather rarely appropriate in the care of children with atopic eczema.

THE BATH

Parents are often given confusingly contradictory advice about bathing. Some doctors tell them not to bath their child, or at least to do so not more than once weekly. On the other hand, others, like myself, tell them to do so every day and, if possible, twice daily. What is one to make of such conflictions?

There is an explanation for the confusion. There can be no doubt that a soapy bath will be harmful to eczematous skin; for this reason, 'normal' baths should unquestionably be avoided.

However, with the availability of modern therapies, the bath can be used as a positive aid to treatment. In fact, I would go as far as to say that it is very difficult to treat atopic eczema successfully without using the bath, and there are in practice very few children who cannot be helped by it.

The functions of bath therapy are several. It helps by:

1. *Cleaning the skin and preventing infection*. The bath should be used to remove the scales, crusts, dried blood, and plain dirt that accumulate on the skin of eczematous children. These are the main attraction for the bacteria that lead to skin infections.

This is why it is particularly important to use a physical skin cleansing routine in the bath, which I will describe later. Such a routine will make the skin surface a less hospitable place for bacteria, and will keep their numbers down. Bathing is the single most important element in your fight against skin infections.

2. *Moisturizing the skin.* Perhaps the most obvious function of the bath is to reintroduce moisture into the skin. The skin of children with atopic eczema is almost always dry, and several factors contribute to this dryness. The skin of children is always relatively dry in any case, mainly because preadolescent children do not yet secrete the oil called sebum on to the surface of their skin (p. 2). Secondly, many children with eczema have one or both of two conditions called *ichthyosis* (p. 26) and keratosis pilaris (p. 19), in which the skin is particularly dry, and which appear to predispose them to have eczema in the first place. And, thirdly, extensive dryness of the skin is caused by the eczema itself, and is itself a cause of itchiness.

The discomfort caused by dryness of the skin is a result of the escape by evaporation of water from the living part of the epidermis (p. 1) through a defective *stratum corneum*. The resulting drying of the living part of the epidermis causes the skin to tighten, which provokes a very unpleasant sensation. The treatment is to replace the lost water, and to seal it into the skin, by applying a relatively impermeable covering of oil over the surface. The bath provides water very rapidly, and the treatments used in the bath help provide the sealing layer.

3. *Preparing the skin for other treatments.* Experience has shown that skin treatments such as topical steroids (p. 109) work much better if they are applied after the skin has been *hydrated*, i.e. its water content has been increased. This hydration is best achieved in the bath. In fact, one of the commonest causes for the failure of topical steroid treatment is that the application has been made on dry, unbathed skin.

The reasons for this failure are probably several. First, the applied treatment is probably mostly soaked up into the dry scales and crusts as if into a sponge, becoming locked up and therefore unable to penetrate into the deeper parts of the skin where its effect would normally be brought about. Secondly, penetration of the steroid itself is slowed down markedly by dryness of the epidermis, and accelerated by hydration. Unbathed skin seems to be as impermeable as leather.

4. *Providing an event for parent and child to enjoy together.* Normal children enjoy their baths, and they provide an entertainment that parents and child can enjoy together. It is one of those 'special' times, which parents normally spend exclusively with a child. If a child is not bathed, the chance of these happy few minutes is lost. Almost all small children love playing in water, and it is often a problem getting them out of the bath.

Children with eczema can find that the bath hurts when their eczema is bad, and it is often abandoned permanently on such an occasion. However, if the bath routine that I am about to describe is used, and other treatments are given along the lines of what follows in this chapter, the bath can be reintroduced into the routine of a child with eczema, however severe, and it can once again become a time of pleasure and a soothing, therapeutic experience as well.

THE BATH ROUTINE

There are three main components to the bath routine for children with eczema. They are:

Add a special bath oil to the water

Adding things to the bath has a long history as a treatment for skin diseases. Years ago, a popular remedy for eczema was to run the water into the bath through a muslin bag containing either oats or bran. The resulting bath was considered very soothing to inflamed skin. There is still available a revived version of this approach in the form of *Aveeno®, which consists of a powdered extract of oats in a convenient sachet, to be added to the bathwater. Two forms of *Aveeno® are made, one with added oil (*Aveeno Oilated®), and one without (*Aveeno Regular®). These preparations suit some children very well. They are definitely worth a try, particularly if your child doesn't seem to get on well with other bath treatments.

Today, adding special oils to the bath is more fashionable and generally more effective. In practice, it is not only an important part of the treatment package your child needs, but also probably the simplest of all treatments for atopic eczema.

There is now available a very wide variety of bath oils made predominantly for use in those with eczema. Some are of the spreading type, some are of the dispersing type, and many are of a combined type (p. 92). The most widely used preparations are shown in *Table 1*.

Table 1 Bath oils

Name	Oil type	Action	Comments
*Alpha Keri®	Liquid paraffin	Dispersing	Contains lanolin
*Balmandol®	Almond	Spreading	
*Balneum®	Soya	Dispersing	
*Balneum Plus®	Soya	Dispersing	Contains an ingredient with anaesthetic properties
*Bath E45®	Liquid paraffin	Dispersing and spreading	
*Diprobath®	Liquid paraffin	Dispersing	
*Emulsiderm®	Liquid paraffin	Dispersing	Contains an antiseptic
*Hydromol®	Liquid paraffin	Dispersing	
*Oilatum Emollient	Liquid paraffin	Dispersing	Contains lanolin

These bath oils are based either on vegetable oils (soya or almond) or mineral oils (liquid paraffin), and may or may not contain lanolin. *Table 1* tends to disguise the fact that there are subtle but real differences in the 'feel' of these oils in use. The fact that two of them contain lanolin should not be considered a disadvantage as so few children are allergic to lanolin. In most cases, between 15–30 ml should be poured into the bath as it is being run.

The more traditional way of doing much the same thing as a bath oil does today was to disperse a preparation called *emulsifying ointment* in the bath water. Emulsifying ointment consists of a mixture of white soft paraffin, liquid paraffin, and emulsifying agents that allow it to be dispersed in water. It does this much more readily in hot than in warm or cold water, and the best method is to disperse it *before* adding it to the bath water. This is best done by putting two or more heaped tablespoonfuls in a heatproof bowl or jug, and then adding boiling water straight from the kettle. Whisking will produce a creamy liquid which should be poured into the running bath. Many dermatologists still recommend emulsifying ointment, and it does have the distinct advantage of being cheap. It has in most ways been superseded by the new bath oils, which are more convenient to use and less messy, but some patients will nevertheless find that emulsifying ointment suits them well and it is worth giving it a try at some stage.

Avoid the use of soap of any sort

Soap is basically an emulsifying agent, and its purpose when used for washing the skin is to remove the mixture of excess oil and dirt that accumulates on the skin during normal activity. Because preadolescent children do not secrete the natural oil called *sebum* (p. 2), their skin tends to be dry normally. For this reason, the use of soap should really be avoided in any child, but it is much more important not to use it in children who have eczema. Soap is the final insult to the irritated dry skin of the eczematous child.

Though some manufacturers sometimes suggest that the problem with soaps for those with eczema is that they contain perfume, this is not the main problem in any case. Some soaps are certainly not as bad as others, either because they are not alkaline like most soaps, or because they have added fat. However, they should nevertheless all be avoided; there is no such thing as a soap that is good for a child with eczema.

Use a suitable cream to clean the skin directly

The main purpose of the bath is to keep the skin clean, and this requires some physical washing activity. This is best achieved by gently massaging a suitable cream into the skin, then rinsing it off (*Figure 43*). This will remove dead skin, crusts, and other debris from the skin

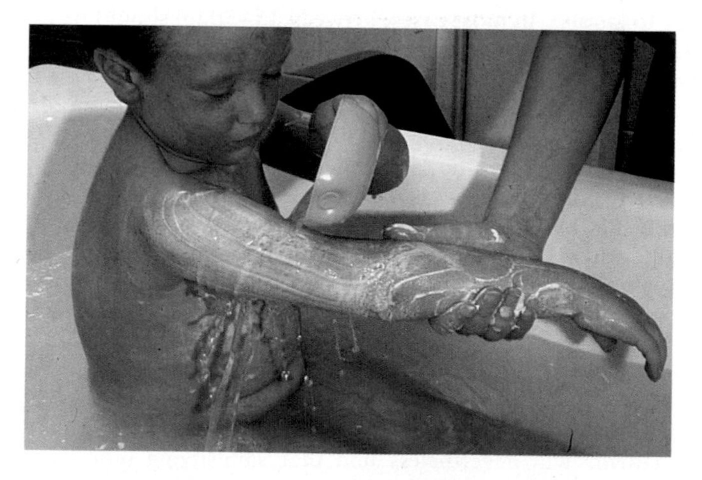

Figure 43 Use cream in place of soap in the bath.

Table 2 Cleansing agents for use in the bath or shower

*Diprobase cream®
*Unguentum Merck®
*Wash E45®
*Aqueous cream BP
*Emulsifying ointment

surface, making it feel more comfortable as well as very much less attractive to bacteria and yeasts. The best technique is to wash the face, the body, and each limb separately in turn.

A wide variety of creams may be used in this way as substitutes for soap, but what they all have in common is that they will disperse in water. The ones I find best are listed in Table 2.

Diprobase cream® and Unguentum Merck® have the great advantage over aqueous cream and emulsifying ointment that they can be obtained in very convenient pump dispensers, though you may have to ask your GP specially for these. The pumps provide larger quantities than would normally be present in tubes, but the biggest advantage is that one doesn't have to dig one's hands in to get the cream out, which carries the risk of contamination. Wash E45® comes in a handy little bottle. Although I have listed them in my own personal order of preference, you will need to find out for yourselves which of these cleansing agents suits your child best.

If your child prefers a shower, these cleansing agents can be used just as easily as in a bath.

Make the bath fun

With small children, it is a good idea to use a doll to get into the bath first, and to have cream applied for washing (*Figure 44*). Have bath toys as for any other child to help keep bathtime fun as well as an opportunity for treatment.

MOISTURIZERS

Eczematous skin is often dry and fissured. The protective stratum corneum (p. 2) is damaged and cracked, and the result is rapid loss of water by evaporation from the living part of the skin. This is a very uncomfortable process, and healing cannot take place while it continues.

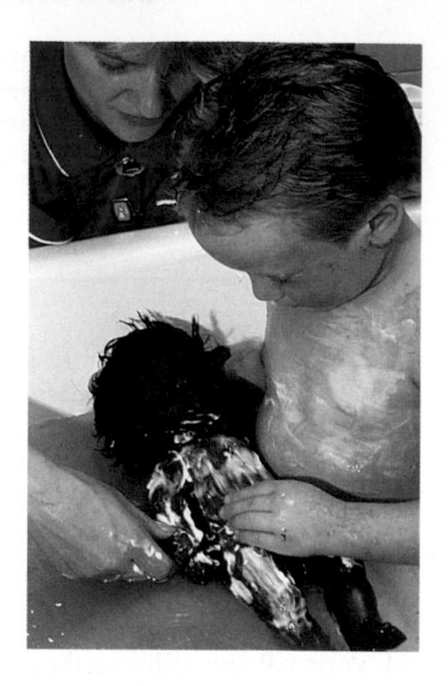

Figure 44 Treating the doll's eczema.

Moisturizing applications all work in the same way, by providing a surface film of oil which slows down the escape of water. This layer of oil also helps by providing a barrier, which will confer some protection against external irritants of the type we considered in Chapter 5.

Moisturizers are more often called *emollients* by dermatologists. A wide variety is available, and no child with eczema is having adequate treatment unless one or other of these is being applied regularly. Those in widest use are listed in *Table 3*.

All of these can be purchased across the counter, and most of them can be prescribed. They vary in a number of ways, but principally in the proportion of water to oil. Adults who do not have eczema would tend naturally to prefer one of the less oily preparations, because they leave the skin less greasy. However, this type of moisturizer will have several disadvantages, which more than outweigh the advantage when it comes to treating a child with eczema.

The less oily preparations inevitably contain more water. Keeping the oil in emulsified form, and preventing bacterial and fungal growth in these preparations, means that they must contain a higher proportion of additives, particularly emulsifying agents and preservatives. These may have an irritant effect on eczematous skin, and, as we have discussed previously (p. 94) many of them are capable, albeit rather

Table 3 Moisturizers

	Preservative	Perfume	Lanolin
Least oily			
*Keri lotion®	+	+	+
A little oilier			
*Aqueous cream BP	+	–	–
*Aveeno cream®	–	–	–
*Diprobase cream®	+	–	–
*E45 cream®	+	–	+
*Hydromol cream®	+	–	–
*Oilatum cream®	+	+	+
*Ultrabase cream®	+	+	–
More oily			
*Unguentum Merck®	+	–	–
*Aquadrate®	–	–	–
*Lipobase cream®	+	–	–
*Oily cream BP	+	–	+
Most oily			
*Coconut oil BP	–	–	–
*'50/50' (equal parts of white soft paraffin and liquid paraffin)	–	–	–
*Diprobase ointment®	–	–	–
*Neutrogena® Dermatological Cream	+	–	–
*White soft paraffin BP	–	–	–

rarely, of providing allergic reactions of the allergic contact eczema type. In addition to this disadvantage, they are unlikely to have as good a moisturizing effect as more oily preparations, precisely because they do contain less oil. The general rule therefore is that one should use as oily a preparation as possible.

Some moisturizers contain special ingredients which are intended to improve their moisturizing qualities. These special ingredients include lactic acid and urea. Lactic-acid-containing applications (e.g. Lacticare lotion®) are beneficial for certain non-eczematous dry skin disorders such as ichthyosis (p. 26), but tend to be too irritating to use in eczema. Urea is a natural substance, a simple chemical, that is found in blood. It is a waste product, and is present in large amounts in urine. It is added to some moisturizers because it has the twin effects of rapid penetration of the skin and attracting water. Its presence in urine may be one of the reasons why the nappy area is often relatively spared in babies with eczema. Although urea undoubtedly does have moisturizing properties, it also tends to irritate and to sting, a little like salt. However, if your child's eczema is dry rather than inflamed, it may be worth trying a urea-containing moisturizer such as *Aquadrate cream®.

In practice, it can be difficult to identify the ideal moisturizer from the available choice. Each child's needs are somewhat different. Although oilier preparations are theoretically preferable, the oiliest moisturizers, such as '50/50', are often too greasy for use in older children. As adolescence approaches, and the skin begins to become naturally oilier, children will usually be happier with a more 'cosmetic' preparation, which does not leave the skin shiny.

Like all topical applications, moisturizers are best applied immediately after bathing, when the water content of the skin will be greatest. However, ideally they should also be applied frequently throughout the day, every one or two hours if possible, especially on exposed areas such as the face and the hands. This is relatively easy to achieve when your child is a baby, and it is a good idea at this stage to apply moisturizer to all affected areas of the body at each nappy change, as well as to the nappy area itself. When children go to play groups and when they start school, you may find that a member of staff is prepared to help by applying moisturizer at regular times. When your child is old enough, it is a good idea to provide a supply for their own use at school. In school-age children, the absolute minimum should be to apply moisturizer before going to school, on returning home, and at bed-time. One should, however, aim in addition to get them to use moisturizer in morning and afternoon breaks and at lunch-time, so long as their eczema is reasonably accessible, particularly when it is present on the face or hands.

PROBLEMS WITH MOISTURIZERS

Moisturizers, indeed all topical applications, may become contaminated with micro-organisms. Just like foods, they may start to grow mould, but the main threat is that they may provide sustenance for bacteria, including those varieties which can cause skin infections. These are likely to be introduced into the application by contact with your child's or your own fingers. This is much more likely to occur with those preparations that come in an open tub, into which fingers are frequently dipped, and with preparations in which bacteria are able to multiply.

As we discussed earlier, creams and lotions principally contain water, which can support the growth of bacteria and moulds. For this reason, the manufacturers have to add preservatives, which are a deterrent but can only slow down the growth of micro-organisms rather than preventing it altogether. The longer a preparation has been kept the more likely it is to have been 'seeded' with micro-organisms, and the more likely it is that they will have succeeded in multiplying. This multiplication will be further encouraged if the application has been kept in a warm environment, and it has free contact with air, especially if the top is left off. Ointments do not contain water, and are therefore much more resistant to the growth of micro-organisms.

Occasionally a child is brought to see me with the story that he or she is 'allergic' to creams and ointments. In actual fact, it is very rare for children to become genuinely allergic to applications for their eczema. The development of allergy can be tested for by 'patch tests', in which the material in question is applied to the skin under a dressing and removed after 48 hours (p. 95). If the child is allergic, a patch of eczema will appear which will still be there a further 48 hours after the test substance has been removed. This sort of test is almost always negative in eczematous children. What actually seems to happen is that the skin in these children has become intolerant of these preparations, and the reasons why this happens are often unclear. Sometimes, the intolerance is to oil of all types, perhaps because blockage of the openings of sweat glands and hair canals has started to cause problems. On other occasions, the problem appears to be irritation of extremely sensitive skin by additives, which can sometimes be overcome by changing to an appropriate alternative application.

The question of whether topical applications block sweat duct openings deserves further consideration. Parents and patients often worry that oily applications might 'block the pores'. In fact, this seems to be a rather rare problem. Sweat seems able to punch a passage through a film of oil without too much difficulty. The use of greasy skin applications to produce what is perceived as an attractive sheen to skin and hair is a widespread practice in many tropical countries. This tradition would be unlikely to have developed if it commonly resulted in skin problems, or if it interfered with the cooling function of sweating.

However, problems do occasionally arise as a consequence of interference with the function of hair follicle mouths by topical applications. Certain types of oil seem to have the property of causing excessive accumulation of *stratum corneum* (p. 1) at the follicle mouth. This may result in blockage, leading to the development of blackheads and acne-like spots. This is quite commonly seen on the forehead of black people who use oils on the hair, both to produce sheen and to straighten the hair. This condition, known as 'oil acne', 'pomade acne', or 'oil folliculitis' is more likely to occur during adolescence and early adult life, when follicular blockage is in any case more likely to occur. Teenagers and adults with eczema may notice that oilier applications cause spots, and they are then likely to be deterred from using these preparations. However, parents should be aware that this type of complication is rare in younger children.

While I always advise parents to use moisturizers that are as oily as their child can tolerate, I am aware that these can have adverse effects which do not affect the child directly but which nevertheless need to be taken into consideration. The very greasy preparations can harm some kinds of clothes, and can be very difficult to wash out. It is rather characteristic for the oilier moisturizers to weaken elastic; this is perhaps the main reason why children with eczema have underpants or knickers that fall down too easily! The oil can also damage the rubber seals on washing machines. It tends to spread around your home, on to your own clothes and soft furnishings. I mention all this to warn you to apply oily moisturizers very thinly, which can help avoid many of these problems. The rule is thinly and often, rather than thickly and rarely.

CONCLUSIONS

I hope I have managed to convey my personal enthusiasm for moisturizers, which I see as an essential element of the basic treatment for all eczema. They are an important aid to keeping the skin clean and free of infection when used in the bath or shower, and this is perhaps the easiest time to use them for their moisturizing effect. It is important, though, to use them at other times as well, as often as is feasible, and ideally every hour if the degree of dryness of the skin warrants it. You may need to experiment a good deal until you find the ideal moisturizers for your child. As your child gets older, and as the eczema changes in character, you may need to change preparations. However, in general, always try to choose the oiliest preparation that you and your child can tolerate, because these preparations will work best, and the effect of each application will last longest. Ointments will also resist contamination by fungi and bacteria, and will provide the skin with some protection against infection.

CORTICOSTEROIDS

Steroids are a group of natural hormones produced in the body, by a variety of different glands. Those which originate in the central part, or *cortex*, of the adrenal glands are termed *corticosteroids*. The adrenal glands are situated just above the kidneys. From the adrenal cortex, these hormones are secreted into the blood, and then circulated around the body. The adrenal cortex produces several different steroid hormones, some which influence the reproductive system (*androgens* and *oestrogens*) and others that influence the balance of salt and water in the body (*mineralocorticoids*). A third variety are known as *glucocorticoids*, of which the most important is called *cortisol*. These hormones are the ones that are of most interest to us, because their principal effects include a damping-down effect on inflammation (*anti-inflammatory* effect) and a suppressive effect on reactivity of certain parts of the immunological system (*immunosuppressive* effect). They have two other effects, whose main importance is that they are side-effects of treatment with these hormones; these are maintenance of blood pressure in situations of extreme stress by trauma or infection, and an inhibitory effect on bone growth.

Glucocorticoid hormones are used to treat a variety of human diseases, and as they are by far the most important type of steroid used in medical treatment generally, they are often known by the abbreviation, *steroid*, even though it is not a very specific term. They should not be confused, for example, with the *anabolic steroids* taken illicitly by some athletes to increase muscle bulk. The first glucocorticoid steroid to be used to treat diseases in humans was cortisol, the name of which is usually changed to *hydrocortisone* when it is manufactured for use as a treatment. Since then, a range of new synthetic steroids have been made which are more powerful and therefore more effective than hydrocortisone. Of these, the hormones most commonly given by mouth are *prednisolone* and the closely related *prednisone*, and the most commonly applied to the skin include *betamethasone*, *fluocinolone*, and *clobetasone*.

A great deal has been written about the dangers of steroid treatment; the result is widespread confusion, even among doctors, nurses, and pharmacists, and this is particularly true when it comes to topically applied steroids.

ORALLY ADMINISTERED STEROIDS

When steroids were first given by mouth for the treatment of disease, it became clear that they could have potentially serious side-effects. If they are given regularly, especially for periods exceeding a month, the adrenal cortex reacts by closing down its own cortisol production. This is no problem from day to day because adequate amounts of hormone are being provided from outside. The major problem is that the adrenal cortex is normally able to increase its production many times in response to the stress of trauma or serious illnesses, such as infections. Its ability to respond in this way is gradually lost if more than minimal amounts of steroid are given by mouth for more than a week or two. This can mean that, in the event of one of these stresses, there may be a deficiency in the supply of cortisol, resulting in a failure to maintain blood pressure. The result can be sudden collapse, potentially dangerous unless the reason is quickly recognized. For this reason, anyone who has been on oral steroids for more than a month will need an increased dose at times of stress. These will include surgical operations, acute trauma such as road traffic accidents, and any illness which is sufficient to keep a child voluntarily in their bed. It is important to be aware that this problem does not occur with topically applied steroids, except in cases of the most extreme overuse.

The other type of side-effect, not so dangerous but nevertheless highly undesirable, results directly from the actions of larger-than-normal amounts of these hormones on various parts of the body. The most important of these is the effect on bones, which causes a weakening of bone in adults and a suppressive effect on growth in children. After a period, high doses of corticosteroids given by mouth will almost invariably produce at least some degree of slowing of growth. The situation is complicated by the fact that many diseases that need treatment with steroids, such as asthma, themselves interfere with growth, and the stimulatory effect on growth produced by successful treatment may more than compensate for the stunting effect of the treatment. The overall effect can therefore be an actual acceleration of growth. Nevertheless, the net effect of oral steroid treatment will generally be a degree of slowing of growth, though this can be minimized to some extent by giving the lowest possible dose, by giving the corticosteroid as infrequently as possible, where feasible only one dose every two days, and by using this form of treatment for the shortest possible period. Furthermore, small degrees of loss of growth will generally by compensated for by 'catch-up' when the treatment is stopped.

Much is said of the suppressive effect that oral steroids may have on the function of the immunological system. In practice, this suppressive effect is confined to those who are taking doses considerably higher than those generally used for eczema or asthma, and children taking oral steroids appear to cope with infections without problems, with the possible exception to chickenpox. If your child is exposed to chickenpox you should tell your GP, who will decide whether to give an injection of protection antibodies (*gamma globulin*). This is only necessary if your child has not had chickenpox previously. Another issue that needs to be considered is that of immunization. The anxiety is that resistance to the weakened (*attenuated*) viruses used for certain immunizations may be impaired in those who are taking oral steroids. The relevant immunizations are polio, and, in the non-routine group, influenza, hepatitis A, and yellow fever. Current policy is to avoid these immunizations in those taking medium and high dose steroids; your doctor will advise.

These are the main problems associated with the use of oral steroids. In many diseases, they provide enormous benefits which cannot be obtained with any other form of treatment, benefits which far outweigh any possible problems. Indeed, it would be no exaggeration to say that steroids contribute as much to human quality of life as any other drug, and would undoubtedly find themselves among the top five if one had to make a 'desert island' choice. The immense good that can be provided by careful steroid treatment is sometimes ignored by their often ill-informed critics.

In childhood atopic eczema, however, their use is only rather infrequently justified, because the disadvantages will generally outweigh the benefits. The place of oral steroids in treating atopic eczema in children is considered later in this chapter (p. 193).

ACTH

An alternative way of giving steroid treatment is to stimulate the child's own adrenal glands to increase their cortisol production. This can be done by giving injections of *ACTH*, which is short for *Adrenal Cortex Trophic* (that is, stimulatory) ***Hormone***.

Under natural conditions, this hormone is produced by the pituitary gland (located at the base of the brain), in response either to stress or to low blood levels of cortisol. It is then carried in the blood from the pituitary gland to the adrenal cortex, where its effect is to stimulate increased production of cortisol. The use of ACTH to treat atopic eczema is discussed on p. 198.

TOPICALLY APPLIED STEROIDS

Steroids may also be used to treat skin diseases by applying them topically to affected areas. They are able to penetrate from the surface to the deeper layers of the skin, where the eczema reaction occurs. When they first became available, it was anticipated that topically applied steroids would revolutionize the treatment of a variety of skin diseases, particularly eczema. However, after the initial unqualified enthusiasm had settled down, it was clear that their use was limited by a variety of problems. They do, nevertheless, have an important place in the treatment of childhood eczema and, as long as certain precautions are taken, they can be used safely.

The original topical steroid was hydrocortisone, which was first introduced into dermatology in the 1950s. It rapidly revolutionized the treatment of many skin diseases, and proved particularly effective in milder cases of the various types of eczema. The pharmaceutical companies were keen to make topically applied steroids even more effective, and did so by synthesizing more powerful steroid hormones with long names such as fluocinolone acetonide, beclomethasone diproprionate, and clobetasol propionate. Though these preparations have definitely proven more effective in treating stubborn eczema, their use is associated with a number of problems which I will discuss later.

When marketing a new topical steroid, the pharmaceutical company selects a suitable base in which the steroid will achieve adequate penetration of the skin and in which it will be stable. Then a concentration will be chosen which results in the final potency the company desires for the product. A product name will be created, such as 'Betnovate®' or 'Synalar®', which tells one nothing about the steroid it contains, nor anything about the strength of the preparation.

The potency of a steroid application will depend on a number of factors, of which the main ones are the innate potency of the particular hormone used, and its concentration. The properties of the base, or *vehicle*, are also important as they can influence the rate at which the steroid penetrates into the skin. *Table 4* is a list of commonly used steroid applications, which I have divided into five groups on the basis of each preparation's final potency. The assessment of potency of topical steroid preparations is difficult and somewhat imprecise, so that this type of ranking list is necessarily rather approximate.

Occasionally, doctors will request pharmacists to reduce the potency of a steroid application by diluting the manufactured product in a cream or ointment base, so that the final strength is a half, a quarter, or even a tenth of the original. If the base is not the appropriate one, the steroid may become chemically altered, usually resulting in a rapid loss in effectiveness. However, information is available to pharmacists telling them exactly which base, or *diluent*, they should use to dilute any particular steroid preparation.

The process of dilution also involves a risk of introducing contaminating micro-organisms, and because of this, these preparations will usually be given very short use-by dates.

In the past, so few less potent preparations were available that the only way to provide them for patients was by requesting the pharmacist to prepare this dilution. However, several manufacturers now

Table 4 The relative potency of topical steroid preparations

Potency	Preparation
Low potency	Hydrocortisone, or hydrocortisone acetate 0.5%, 1%, or 2.5% Efcortelan® Hydrocortistab® Mildison Lipocream®
Low–medium potency	Clobetasone butyrate 0.5% Eumovate®
	Flurandrenolone 0.0125% Haelan®
Medium potency	Fluocinolone acetonide 0.00625% Synalar 1 in 4® dilution
	Betamethasone valerate 0.025% Betnovate RD®
	Desoxymethasone 0.5% Stiedex LP®
High potency	Betamethasone valerate 0.1% Betnovate®
	Beclomethasone diproprionate 0.025% Propaderm®
	Fluocinolone acetonide 0.025% Synalar®
	Hydrocortisone 17-butyrate 0.1% Locoid®
	Betamethasone dipropionate Diprosone®
	Desoxymethasone 0.25% Stiedex®
	Fluocinonide 0.05% Metosyn®
	Budesonide 0.025% Preferid®
	Diflucortolone valerate 0.1% Nerisone®
	Triamcinolone acetonide 0.1% Aureocort®
Very high potency	Clobetasol propionate 0.05% Dermovate®
	Fluocinolone acetonide 0.2% Synalar Forte®
	Halcinonide 0.1% Halciderm®

make these dilutions themselves under carefully controlled conditions which overcome these problems, and 'home-made' dilutions are much more rarely needed. One situation in which a 'home-made' dilution may be appropriate is in connection with wet-wrap dressings, which will be described later (p. 129). If a dilution of a steroid application has to be made for your child, you should do your best to make sure that your pharmacist uses the correct diluent.

PROBLEMS WITH TOPICALLY APPLIED STEROIDS

There are three principal problems that may arise when steroids are applied directly to the skin. These are problems that result from absorption of the hormone into the blood, resulting in internal effects, problems that arise in the skin itself from excessive exposure, and something called 'tachyphylaxis'.

Internal (*systemic*) side-effects of topically applied steroids

The ability of topically applied steroids to pass through the skin inevitably means that a proportion of it will end up in the blood. If the amount reaching the blood is great enough, the result can be problems identical to those that occur when steroids are given by mouth. The danger of this happening depends on four factors: the potency of the preparation used, the extent of the area treated, the amount applied (clearly this is at least partly related to the area treated), and the age of the child.

There is a definite risk of internal side-effects when Group 4 or 5 preparations are used on eczematous children. Clearly, the risk will be greatest when large amounts are used on extensive areas of skin. The risk will also be greater in smaller children, because the ratio of their surface area to their volume is relatively higher. This means that the hormone absorbed from treatment of one square centimetre of a baby will have a relatively greater effect on the whole child than it would on a bigger child. It is impossible to give precise guidelines as to what is or is not safe. Group 1 preparations *virtually never* cause internal side-effects unless they are used extremely extravagantly in very small babies. To have an effect of this type, they would have to be applied to most of the child's surface area at least twice daily. Group 2 preparations are unlikely to cause internal side-effects except in babies. Group 3 preparations could, however, do so if used on extensive areas in large quantities over a long period, especially in the case of children under the age of five years. No dermatologist would ever now use Group 4 or

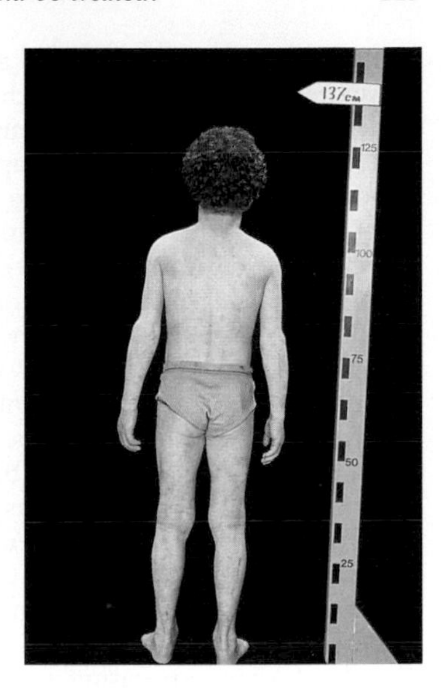

Figure 45 This ten-year-old boy's growth has been slowed down by absorption of strong steroid through the skin over many years. He should reach at least to the marker.

5 preparations undiluted for eczema in children of any age for periods exceeding a few days.

Although it would be nice to have exact guidelines as to what is safe and what is not safe in this regard, this probably isn't realistic because there seems to be a great deal of individual variation as far as susceptibility to these side-effects is concerned. In other words, what is perfectly safe for one child appears not to be in another. In practice, the main problem if excessive quantities of steroid are absorbed will be slowing down of growth (*Figure 45*). I have known children to be harmed in this way by what many dermatologists would regard as perfectly acceptable amounts of steroid use. The only way to make absolutely sure that this type of side-effect is avoided is from time to time to measure the growth of children treated with topically applied steroids.

External (*local*) side-effects of topically applied steroids

Steroids may also directly damage the skin. This type of damage is called *atrophy*, and though it occurs to a degree when steroids are given by mouth, it tends to be more of a problem with topically applied preparations because the concentration reached locally in the skin is generally higher. Steroid-induced atrophy is very similar to what happens to the skin quite naturally as a result of ageing. It becomes thin, transparent, and fragile. These changes are due to thinning of both the epidermis and the dermis, though the effect on the latter is more profound. As mentioned in Chapter 1, the bulk of the dermis is provided by a mass of intertwining fibres made of a protein called collagen. In the normal skin, production of new collagen fibres goes on alongside dissolution of old or damaged fibres. Both in old age and under the effect of corticosteroids, the balance of production and dissolution is altered so that new production is inadequate to replace losses and the dermis gradually wastes away. The epidermis loses its support and becomes more fragile. The blood vessels that run in and through the dermis become more easily visible from the surface, and a lack of support of the smaller blood vessels will mean that they are more easily torn, resulting in easy bruising.

Early on, this atrophy is reversible; if steroid treatment is stopped, the skin will return more or less completely to normal. The dermis can, however, become so badly damaged that it loses its elasticity altogether; at this stage 'stretch marks' (*striae distensae*) will appear, just like those often appearing on the abdomen in pregnancy. If this happens, the skin will never again return completely to normal; striae are permanent.

The rapidity with which atrophy develops depends on the potency of the preparation, and the frequency with which it is applied. It also depends on the site. Striae are particularly likely to occur on the insides of the thighs (*Figure 46*), the insides of the upper arms, and on the abdomen or breasts. However, these are also sites where they may develop for other reasons, particularly as a result of rapid growth during puberty, as a result of tension in the skin during pregnancy, or because of obesity.

One of the most obvious effects of topically applied steroids on the skin is a blanching effect, caused by constriction of small blood vessels. However, this constriction tends to be followed by a rebound period of opening up of the blood vessels (*vasodilatation*), which is initially temporary but which can eventually become permanent. This results in even more apparent small blood vessels in the skin (known technically as *telangiectasia*). The face appears to be especially susceptible to this side-effect of stronger topically applied steroids. For this reason, preparations other than those in Group 1, or occasionally Group 2, should , generally speaking, not be applied to the face.

Steroid-induced skin atrophy and telangiectasia take several weeks to develop, and will therefore be avoided if potent preparations are used for no longer than a week at a time. It is therefore perfectly

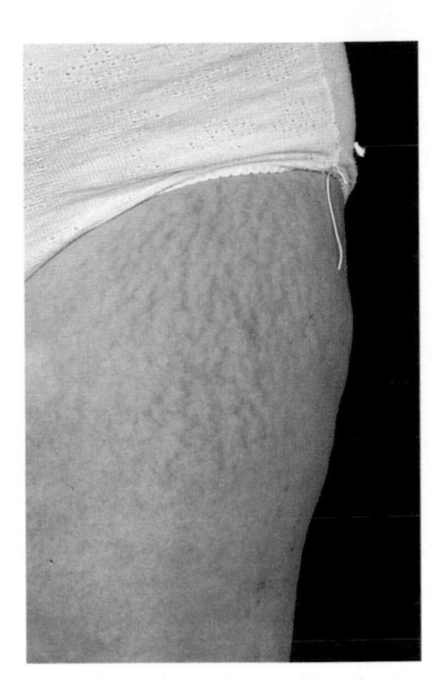

Figure 46 Stretch marks from use of an inappropriately strong steroid cream on the thigh.

justifiable to apply these preparations for a few days as an emergency measure to cope with a sudden flare-up. Many dermatologists use schedules during which a Group 4 or a Group 5 steroid is applied for a few days during an exacerbation of eczema, followed by application, each for a few days, of decreasingly potent preparations in order to try to avoid a rebound deterioration. This is perfectly safe as long as there is no danger of continuing to apply the more potent preparations, and no danger that the schedule will be repeated too soon.

Group 4 and Group 5 preparations should only in exceptional circumstances be used for longer than one week at a time, with at least a four-week interval between such courses. Group 2 and 3 preparations will not cause visible atrophy except on the face.

Hydrocortisone alone (Group 1) *never* causes atrophy, even on the face. It must be apparent by now that hydrocortisone is unique in its margin of safety and, although there may be a place for more potent corticosteroids in older children, nothing stronger than hydrocortisone should *ever* be used in babies.

One further undesirable effect of topically applied steroids is increased pressure in the eye, known as *glaucoma*. This is rather rare, but may occur if a preparation from Group 3, 4, or 5 is applied persistently to the eyelids, because of penetration through the lid on to the surface of the eye, and, from there, through into the eyeball itself.

TACHYPHYLAXIS

Tachyphylaxis is the medical term for a phenomenon that occurs with topically applied steroids (but, curiously enough, apparently not with steroids given by mouth). After a period of regular use, any particular preparation will generally gradually start to lose its effect. How rapidly this occurs depends largely on how often the preparation is used.

When a particular preparation is applied daily, this loss of effect is almost invariable. If application frequency is greater than twice daily, it may already appear after only a few days, and it will become gradually more and more noticeable thereafter. It will occur more slowly with less frequent daily application, but may not occur at all if application is only occasional. The effect also seems to happen more rapidly with more potent preparations, particularly those in Group 4 and Group 5. It can be partially overcome by using it less frequently or, where this is not feasible, by changing to another preparation. Often a good tactic is to change brands within the same potency group every few weeks. After a particular preparation is stopped for a few weeks, it will often be found that it has regained much of its former potency: in other words, 'tachyphylaxis' is reversible. Unfortunately, with the more potent Group 4 and Group 5 preparations, it is found that the effectiveness of the whole group wanes with continuous use, even if preparations are changed regularly.

The effect of steroid applications can be amplified by the use of polythene film. This technique is known as 'occlusion', and it works by increasing hydration of the skin. Hydration enhances the transport of steroids through the barrier of the stratum corneum, and occlusion will therefore accelerate the rate at which atrophy occurs when potent steroid preparations are used. The widespread use of occlusion to treat eczema and other skin diseases during the 1960s and 1970s was responsible for much of the adverse publicity about steroid side-effects which continues today.

Polythene occlusion of steroids is still occasionally used, however, to treat hand or foot eczema in adults, and when done with care and skill can be very effective. This technique is now almost never used in children with atopic eczema because the areas which normally require treatment in the severely affected case are likely to be extensive. However, there are times when modifications of this approach may be useful to deal with limited areas of persistent eczema, even in children. One company produces a plastic tape impregnated with a steroid, which although otherwise mild may have a very powerful effect when used in this way (Haelan Tape®).

Another way of achieving a temporarily more convincing effect from steroid treatment is to use a partially occlusive technique with one of the modern semi-permeable film dressings. A favourite of mine is occasionally to use a dressing called *Tegaderm®. These semi-permeable films allow air and water vapour to pass through but only slowly. Applied over a suitable steroid application they will increase its effectiveness, and can be used to clear eczema from sites of limited extent in which it has become very persistent. The use of these semi-permeable films is much safer and more comfortable than polythene film, as the occlusion is not so complete. As long as this approach is used only as a temporary measure, it can be both successful and harmless.

A lesser degree of hydration is, of course, achieved by bathing and by using moisturizers, both of which also enhance the permeability of the stratum corneum to steroids applied subsequently. This is an effect on which we depend for steroids to penetrate this barrier, and without this degree of hydration one may get virtually no effect whatsoever from the treatment. In fact, all steroid preparations are made up in what is effectively a moisturizer. As we considered earlier (p. 101), ointment bases are better hydrators than creams and lotions and therefore work better. This is also the reason why steroids are best applied after the bath.

Perhaps the most common reason for failure of applied steroids to have much effect on eczema is that they have been applied to un-hydrated skin. If steroids are applied to dry scaly areas of eczema, penetration may be so poor that they fail to produce any benefit at all.

In recent years it has become clear that one of the most successful ways of all to use topically applied steroids is under wet dressings often known as wet wraps. I will describe this very valuable technique in detail later on (p. 129).

TARS

Tars are obtained by distillation from several sources, including coal and wood. It has long been recognized that tars can have a soothing effect on inflamed skin, and they are a traditional remedy in many countries for several skin diseases including eczema. Tars contain hundreds of chemicals, most of which have never been identified, and many of which have medicinal effects.

The principal tars for treatment of skin disease in the UK come either from coal (coal tar) or from shale containing fossilized fish (ichthammol).

COAL TAR

At one time it was common for crude coal tar to be applied directly to the skin. Unfortunately, this could itself be irritating; it is also messy, smelly, and, cosmetically speaking, a near disaster. It is therefore more usual these days to apply it in a diluted form known as coal-tar solution. This can be added to a cream, ointment, or paste base, or can be added to the bath. Some very mild preparations combining tar have become available in recent years; a good example is *Clinitar® cream. Some of these, such as ’Tarcortin® cream, contain hydrocortisone to increase the anti-inflammatory effect.

ICHTHAMMOL

Ichthammol (from *ichthyos*: a fish, in Greek) is much milder than coal tars, and is in many ways a more suitable type of tar for treating eczema. Possibly the best way of using any tar, but particularly ichthammol, is in the form of paste bandages, which I will discuss in a later section of this chapter.

HERBAL AND OTHER NATURAL AGENTS FOR APPLICATION TO THE SKIN

Calamine is a naturally occurring mineral, zinc carbonate, coloured pink by a trace of iron oxide. Calamine lotion is a traditional soothing application which has found its main use in burns, sunburn, insect stings, and so on. Calamine lotion has a pronounced drying effect and is therefore inappropriate for eczema in dry phases. It may be used for exudative eczema, but tends to cause a build-up of lumpy crusts, which are unsightly and delay healing. Recently, calamine has been incorporated into a cream called Eczederm®, which does not have a drying effect and may be used as a soothing moisturizer in milder eczema. Calamine is also an ingredient of some paste bandages (p. 138).

Chamaemelum nobile (Roman Camomile) is a plant with a long history of use in Europe as a medicinal herb. It is considered soothing when applied directly to the skin. Interestingly, it is available in the form of *Kamillosan® ointment, and this is well worth trying as a treatment for mildish atopic eczema.

Calendula officianalis (Pot Marigold) is another herb with traditional medicinal use in Europe, and is often prescribed in cream or ointment by homeopathic practitioners and by herbalists. Like camomile it is easily available, as *Calendolon® ointment, and may similarly be worth trying for mild eczema.

OTHER SKIN APPLICATIONS

A variety of other agents are sometimes prescribed for application to eczematous skin. These include crotamiton, calamine, so-called non-steroidal anti-inflammatory agents, antihistamines, and local anaesthetics, though none of these are of established benefit.

Crotamiton is an agent which has mild anti-itch properties, and is available as a lotion or ointment (*Eurax®), or combined with hydrocortisone in a cream ('Eurax Hydrocortisone®). In practice, these preparations are of little if any value in the treatment of eczema, and they may have an irritant effect if the eczema is in a weeping phase.

Non-steroidal anti-inflammatory agents are drugs that have been developed as an alternative to steroids, mainly for treating arthritis. Well-known examples of these include indomethacin and phenylbutazone. Though there is no evidence that these agents are effective for eczema when given in their usual oral form, similar drugs have been incorporated into preparations for topical application to the skin. The most easily available of these is bufexamac ('Parfenac® cream). This cream is widely used to treat atopic eczema in other countries, but good evidence that it is effective is lacking. It appears to be safe, though occasionally contact allergic reactions have occurred. It may be worth a try.

Antihistamines are undoubtedly of value as an orally administered treatment for children with atopic eczema (p. 149). These drugs have also been incorporated into various lotions and creams for direct application to the skin. The most widely used of these preparations are *Caladryl® cream and lotion, which contain both an antihistamine and calamine. These can be purchased over the counter at chemists, and parents may be tempted to buy them. Though there is little evidence that they are effective at relieving itching, this type of preparation is used quite widely as a treatment for sunburn and insect bites. Their downfall is a distinct tendency to provoke contact allergic reactions after they have been used for any length of time, and the development of this type of allergy is serious because it means that person will have become allergic to antihistamines and closely related drugs given by mouth as well. This kind of allergy is virtually permanent once it has arisen, and its presence can be quite a serious problem because drugs of this type are used to treat not only eczema but also asthma, hayfever, rhinitis, urticaria, and a wide variety of psychiatric disorders, including anxiety states. I would recommend very strongly that this type of agent should not be used, for any purpose, in children with eczema.

Local anaesthetics can temporarily relieve the sensation of itching, but to maintain this effect requires virtually continuous application, which in the long term is found almost inevitably to lead to contact allergic reactions. If this happens, there will be danger when local anaesthetics are injected (for dental work, for example). It is perhaps unfortunate that these preparations can be obtained over the counter, though generally the accompanying leaflet will state that their use is contraindicated in eczema.

ANTIBACTERIALS

In Chapter 7, I described how eczematous skin inevitably supports an unusually high population of bacteria. When their numbers are low, they appear to do little harm, a situation we call *colonization*. On the other hand, some varieties of bacteria are capable of making the eczema worse, should their population reach a sufficient density; this we would call infection. The variety known as *Staphylococcus aureus* is responsible for the great majority of significant bacterial infections in children with eczema.

In practice, the problem is to distinguish between colonization alone and infection proper. Currently there is no simple and widely available test to enable us to quantify the bacteria on the skin, so that we largely have to rely on visual clues to detect infection. We would suspect infection wherever there is rapid worsening of eczema with the development of weeping, or pustules and subsequently of crusts with a particular yellow colour (*aureus*: golden, in Latin). These changes in the skin are often combined with painful enlargement of lymph nodes (p. 205) and, when the infection is more profound, of malaise and/or fever. The problem is that only the more severe infections reveal themselves in these ways. Many infections appear to cause nothing more obvious than worsening of the eczema. Infection should therefore always be considered whenever your child's eczema worsens unexpectedly, and is in fact probably the commonest cause of unexplained deterioration in atopic eczema. Because we lack a reliable laboratory test for infection, the detection of infection requires alertness.

The most important treatment for skin infections in eczema is prevention. Effective prevention involves making the skin as unattractive as possible from the point of view of bacteria. This means good skin care, particularly frequent baths or showers, using appropriate oils in the water, and creams to clean the skin physically in place of soap, as discussed earlier in this chapter. It also means the frequent application of moisturizers at other times, to seal the surface of the skin with a film of oil, which is very much to the distaste of bacteria. Antiseptics have very little place in this respect, because they are generally irritating to the skin unless they are used so dilutely that they are ineffective.

INTERNALLY ADMINISTERED ANTIBIOTICS

The internal (*systemic*) use of antibiotics in eczema is generally reserved for obvious and extensive infections, or for exacerbations in which one strongly suspects that infection is responsible. Since the great majority of infections are caused by *Staphylococcus aureus*, the initial choice of antibiotic will be one that will deal with this bacterium. In the UK, flucloxacillin ('Floxapen®) is generally the preferred antibiotic, but in other countries a very similar antibiotic called cloxacillin ('Orbenin®) may be used. These antibiotics are closely related to penicillin, but have been modified to make them more effective against *Staphylococcus aureus*, which very often resists treatment with penicillin itself. A good alternative is erythromycin ('Erythroped®, 'Ilosone®). Erythromycin is safe, cheap, and effective in the majority of infections in eczematous skin. Other antibiotics that are effective for *Staphylococcus aureus* infections include sodium fusidate ('Fucidin®), clindamycin ('Dalacin C®), cefaclor ('Distaclor®), cephalexin ('Ceporex®, 'Keflex®), cefadroxil ('Baxan®), trimethoprim ('Monotrim®, 'Trimopan®), and cotrimoxazole ('Septrin®, 'Bactrim®).

It can be useful, but is not always necessary, for a doctor to take a swab before treating infected eczema with an antibiotic. This will provide information about the type of bacteria present, which would be valuable if the infection failed to respond to the initial treatment. Furthermore, a range of antibiotics can be put on to the culture plates and their effect on the growth of the bacterial colonies can be observed. This will provide information about their susceptibility to different antibiotics, and will also be valuable if the infection is not cleared up by the antibiotic first prescribed.

The second most common type of bacterium to cause infection of eczema is called Group A streptococcus. The bacterium causes an infection that often looks different from infection with *Staphylococcus aureus*. There tends to be less pus and crusting, but more redness, tenderness, and swelling. This bacterium frequently causes sore throats, and a clue will be if the eczematous child or another family member has a sore throat or tonsillitis. If this bacterium is considered likely to be responsible for an infection, another antibiotic, usually either phenoxymethylpenicillin ('Distaquaine V-K®, 'Stabilin V-K®, 'V-Cil-K®), ampicillin ('Penbritin®), or amoxycillin ('Amoxil®) may be given, either alone or in combination with one of the antibiotics listed above.

Antibiotics are always prescribed for a set period, usually for three, five, seven, or ten days. Even though the infection for which they were given may seem to be effectively treated after two or three days, they should *always* be taken for the full period. This is to try to ensure that as many bacteria as possible are killed before stopping treatment. If the proportion killed is low, the infection is likely to become re-established quickly, with the additional problem that the bacteria may now have acquired resistance, precisely because they survived the previous inadequate treatment. So always take antibiotics for the full period.

With all antibiotics, success depends on the maintenance levels in the blood throughout the 24 hours of each day. If the blood level does fall below a critical level, it allows the bacteria to proliferate afresh and this delays successful treatment. It may also allow bacteria to become resistant during the period when the concentration is low, and this may prevent successful eradication of infection. 'Take three times daily' therefore means that the antibiotics should be given more or less exactly every eight hours. Antibiotics that are supposed to be given 'four times daily' should be given every six hours, though this will be difficult if your child sleeps for eight hours or more. In this case you should aim to give the treatment as your child goes to bed, and first thing on getting up in the morning, spreading out the other doses evenly during the rest of the day.

Resistance tends to develop in bacteria if they are exposed for any length of time to amounts of an antibiotic less than that required to kill them. No antibiotic is lethal to all types of bacteria and some will survive treatment. Our intestines are full of bacteria that help digestion, and which we need to remain healthy. If a systemic antibiotic is used to treat an infection in the skin, many of these gut bacteria will survive. Some of them will develop resistance to the antibiotic used, and can sometimes pass it on to other potentially harmful types of bacteria. Fortunately, this is usually only temporary, and the resistance usually disappears after a period unless the same antibiotic is used again. The problem with resistant bacteria is a small one in the home, but can be quite a big problem in hospitals where antibiotics are used frequently. As patients are usually in close contact with one another, resistance can spread and this is one reason why children with infected eczema are often nursed in separate rooms away from other patients.

The principal adverse effect of internally administered antibiotics is allergic reactions. These may take many forms, and can occasionally be serious. The more often one is given a particular antibiotic, the more likely one is to develop an allergy; this is one of the main reasons for giving antibiotics as infrequently as possible. Certain antibiotics may occasionally cause diarrhoea, which may be due to destruction of the bacteria that normally inhabit the large intestine, but other side-effects are rather uncommon.

TOPICALLY ADMINISTERED ANTIBIOTICS

Antibiotics are sometimes prescribed for topical application to infected eczema, rather than being given internally. This has advantages and disadvantages. The main advantage is that the danger of severe allergic is reduced. The main disadvantage is the difficulty of being sure that the bacteria are adequately exposed to the antibiotic, both because they may be proliferating in sites not being treated, such as inside the nose, and because the concentration of the antibiotic in the skin may not be adequately maintained throughout the 24 hours of the day. Both these factors have the effect of encouraging the development of bacterial antibiotic resistance. For these reasons, the principles involved in the use of topical antibiotics are different.

First, they should be used for severe or extensive skin infections only if an antibiotic is also given internally. Secondly, drugs used for serious internal infections such as lung or kidney infections should never be used on the skin. This limitation applies particularly to gentamicin ('Genticin®).

A further problem with topical use of antibiotics is that in some cases contact allergy may develop very quickly; this is particularly true of penicillin and related drugs, so much so that they are never used in this way.

Some antibiotics which are either too toxic to be given internally or are not absorbed into the body from the intestines, can nevertheless be used safely on the skin. Examples of the latter include mupirocin ('Bactroban®), clioquinol ('Vioform-HC®) and chlorquinaldol ('Locoid C®). Tetracycline may stain the teeth if used in children under twelve, but it can be used safely topically ('Terracortil®, 'Trimovate®, 'Aureomycin®, 'Achromycin®).

Neomycin ('Cicatrin®, 'Graneodin®) is sometimes prescribed for topical use. This drug can cause deafness if enough enters the blood, and this undoubtedly can happen if it is used on extensive areas of broken skin. I would therefore recommend that it should be avoided for treatment of eczema. The same applies to other topical antibiotic preparations, including 'Soframycin®.

The way that doctors use these topical antibiotics in eczema varies a great deal. Some reserve them for obvious infections, others use them more readily—even when infection is not clearly present. My own feeling is that they generally have very little value in treating children with eczema except in certain situations, specifically very localized infections and treatment of nasal carriage of *Staphylococcus aureus*.

Sometimes a child with eczema will have a problem with very localized infections which do not always justify treatment with an internally administered antibiotic. A good example would be the infection under the free edge of a nail which seems to result from an accumulation of dead skin and other debris under the nail as a result of scratching. A topically applied antibiotic may be very effective for this type of infection. Some children have problems with recurrent infections which may always appear to start as a very localized group of pustules, for example, and it may be possible to 'nip the infections in the bud' by providing a topical antibiotic for immediate use. The best topical antibiotics for these forms of infection appear to be mupirocin ('Bactroban®) and fusidate ('Fucidin®).

Treatment of nasal carriage is considered below.

TREATMENT OF RECURRING SKIN INFECTIONS

In some children, a new infection seems to develop just as soon as the last one has been treated. Most of the campaign against recurring infection must take the form of prevention, as mentioned earlier, by making the skin as unattractive as possible to the bacteria that cause these infections. However, despite these preventive measures, some children seem to continue to be prone to reinfection. A deficiency of the immunological system is sometimes suggested by doctors to explain such a susceptibility, but careful investigation almost always reveals that the child has a fully functional immunological system, and that there must be some other explanation.

Staphylococcus aureus are not all identical, and different subtypes or strains can be identified. Experience suggests that the bacteria causing recurrent infection of a particular child's eczema will usually be of the same strain as those that caused the previous infection. They do not therefore seem to be a result of transmission of new infections from other people. The main problem seems to be that conditions in the skin are so favourable to the small numbers of bacteria that manage to survive each period of antibiotic treatment that they rapidly multiply until the population is large enough to cause yet another infection.

Many of those who are physically close to a child with infected eczema will themselves become colonized by the child's strain of *Staphylococcus aureus*. The usual site for such colonization in those with healthy skin is the inside of the nose, from where the bacteria can easily get around on the fingers. Sometimes, family members will themselves get full-blown skin infections, usually in the form of impetigo, crops of pus-filled spots, boils, or infections in cuts, abrasions, and so on. It is likely that on at least some occasions the bacteria causing recurrent infection have been transmitted back to the patient in question from a close relative or friend who had themselves become colonized or infected in this way. This could happen despite total eradication of the initial infection in the patient. The implication is that one needs to consider elimination of colonizing *Staphylococcus aureus* in close relatives and friends, as well as obvious infections. Therefore, if children with eczema are getting recurrent infections despite good skin care and adequate antibiotic treatment of infections as they occur, one measure one has to consider is local antibiotic treatment of the nose in close relatives and possibly also in close friends. This is best achieved with 'Bactroban Nasal® ointment or 'Naseptin® cream, applied twice daily for seven days to the accessible part of the inside of each nostril. In both children and adults it is a good idea also to apply the same ointment to the finger tips and under the fingernails. I often wonder whether family pets might not occasionally be the source of reinfection, as dogs and cats often have skin wounds and are as liable to nasal colonization as humans. When this seems worth worrying about, a discussion with the vet would be advisable.

Another measure that needs to be considered where recurrent infections are a problem is a period of continuous preventive (*prophylactic*) antibiotic treatment. Generally, a period of three months' treatment will be sufficient, and the appropriate dose will be about a half of the normal full dose. The antibiotic to be used for this purpose will need to be carefully selected in order to minimize the risk of resistance developing that would threaten successful treatment of fully developed skin infections in the future.

Where all these measures fail—good skin care, attention to colonized relatives, and a course of prophylactyic antibiotic—the next thought will be investigation of the child's immunological system. It is important, however, to be aware that the clue to a faulty immunological system is frequent infection also occurring at sites other than the skin such as the ears, sinuses, lungs, and bones. The occurrence of frequent infections at these sites in an eczematous child should lead to their referral for immunological investigation. Generally, however, these types of infections will not be occurring, and the problem will be largely restricted to the skin. In this event, the next move will be for the child's doctors to consider a step up in the level of eczema treatment itself, since the infections will stop if the eczema can be got under good control—by whatever means.

'MY CHILD'S ECZEMA CLEARS UP EVERY TIME HE TAKES ANTIBIOTICS'

Many parents have told me that, when their child is treated for infected eczema, not only the infection but the eczema itself seems to disappear. Such parents often find that this will happen every time their child is treated in this way.

Presumably this observation indicates that the bacteria that caused the infection are responsible to a large degree for the maintenance of the eczema even when no obvious signs of infection are present. This may mean either that infection is present under normal circumstances but is undetected, or that the bacteria are able to worsen the eczema even when they are merely colonizing the child's skin. We have no reliable way at present of distinguishing between these two possibilities. In Chapter 5, we considered the possibility that bacteria such as *Staphylococcus aureus* might cause problems as a result of a person with eczema becoming allergic to them or to their products. It seems highly likely that this might be an important factor in some cases, but conclusive proof is lacking.

Whatever the theoretical explanation for this observation, it clearly raises the question whether it would not be justifiable to attempt to eradicate the responsible bacteria from such children in the longer term. Since topical antiseptics and antibiotics do not appear able to achieve this end, one would have to give a suitable antibiotic by mouth continuously for a period of many months. This may be worth considering in certain cases, though in general one would avoid giving long-term antibiotics for fear of provoking the development of antibiotic-resistant strains which could then cause infections that would be difficult to treat. Clearly, the pros and cons would need to be carefully assessed in the individual case.

ANTISEPTICS

Antiseptics are poisons which kill bacteria, but which are also capable of killing human cells. Because they are poisonous they cannot be used internally. On the other hand, in suitably diluted form they may be safely applied directly to the skin or added to the bath-water. The intensive use of antiseptics is a feature of the treatment of eczema in some other countries, particularly France, and to a lesser extent the USA.

Antiseptic solutions are sometimes used to treat eczema which is weeping. The aim is to dry up and stop the exudation. There are several alternative antiseptics for this type of use; probably best of all are dilute solutions of either potassium permanganate or aluminium acetate (known in the USA as *Burow's solution*). These are made up as required in warm water from either crystals or a concentrate. The ideal proportions are one part of potassium permanganate to 8000 parts of water and one part of aluminium acetate to 500 parts of water. The affected areas of the skin can either be immersed in the warm solution or it can be applied to the skin as a wet compress. Probably the best way of making such compresses at home is to use old boiled nappies or towels. These are soaked in the freshly-made solution and then placed on the area to be treated. Every five minutes or so the compress should be removed, squeezed out, and replenished with more of the warm solution. Which method is chosen will depend largely on the site to be treated. A hand or foot will probably be most easily treated by immersion, whereas the cheeks could be treated only by the wet compress method. Each treatment should last 15–20 minutes and 2–4 treatments are given each day. A problem with potassium permanganate is that it stains the skin (and anything else that comes into contact with it) a brownish colour, but it is nevertheless the favourite antiseptic for this purpose in the UK.

One might expect that the liberal application of antiseptics to the skin would lessen the risk of infections occurring by reducing numbers of colonizing bacteria. Unfortunately, there are two snags that in my view mean that such treatment is best avoided. First, antiseptics tend to be irritant at the concentrations required to kill bacteria on the skin, and, secondly, application of such chemicals to large areas of skin is inevitably accompanied by some absorption into the blood. Since antiseptics are by their nature toxic, this is potentially hazardous with long-term use. Perhaps very occasionally a case can be made for using antiseptic in the bath in a child who is having problems with recurring infections, but there is no convincing evidence that this approach is effective. This should certainly not be attempted in any child under the age of two years because of the dangers from absorption, and only those antiseptics with an acceptable safety profile should be considered. This more or less restricts one to triclosan or potassium permanganate. Triclosan is available as *Sterzac® Bath Concentrate, of which 28.5 ml should be added to a standard 140 litre bath (less if the bath volume is smaller). Potassium permanganate should be used in a concentration of about one part in 8000, which has the appearance of diluted blackcurrant cordial. Even when diluted, domestic antiseptics such as Dettol® tend to be too irritant to use in this way.

EVENING PRIMROSE OIL

In recent years there has been a great deal of interest in the use as a treatment for atopic eczema of the oil obtained from the seeds of the evening primrose (*Oenethera biennis*) (*Figure* 47). This interest goes back just over ten years when the first indications emerged that evening primrose oil could be a valuable treatment when given by mouth. There is some evidence that those with atopic eczema have a deficiency in the blood of a fat known as gamma-linoleic acid. Evening primrose oil is especially rich in this substance. However, despite a number of trials in patients, there is still considerable scientific controversy about beneficial effects of evening primrose oil and the situation remains unclear. I share the view of many of my colleagues that, at least in children, evening primrose oil does not offer a major advance in treatment for atopic eczema. In the recommended dosage, relatively few children appear to experience a worthwhile degree of improvement after up to three months' treatment. However, many of us feel that there is a small number of children who do benefit, and that though there appears to be no way currently of identifying these individual children, evening primrose oil may be worth trying where simple external treatments alone have failed. Evening primrose oil has the additional advantage of being free from any known hazard at the recommended dosage.

Figure 47 The evening primrose, *Oenethera biennis*.

I believe that a three-month trial may be justified in any child with extensive eczema that has not responded to straightforward external treatments. Evening primrose oil is prescribable in the form of *Epogam® or *Epogam Paediatric® capsules. It is often easiest to give children the oil mixed with milk, spread on bread, or added to other food. The advantage of the paediatric capsules is that they have a small extension that can be snipped off with scissors, after which the oil can easily be squeezed out. The dose is eight capsules daily of Epogam®, or four capsules daily of Epogam Paediatric®, for children of all ages over one year. Currently there is little experience of its use in children less than a year of age.

If, after three months, there has been no discernible improvement, the treatment should be discontinued. On the other hand, if it is felt that there has been a worthwhile degree of improvement following a three-month trial, it is worth continuing treatment for a further three months and possibly longer, but the dose can probably be halved. A great deal more research will be required before the role of evening primrose oil, and the most effective dose and length of treatment can be more firmly established.

DRESSINGS USEFUL IN THE TREATMENT OF ECZEMA

Dressings have several uses in the treatment of atopic eczema, and can make a very substantial contribution to effective therapy. I shall consider them in turn, under the following headings: wet wrap dressings, mitts and dry wrap dressings, medicated bandages, and occlusive and semi-occlusive dressings.

WET WRAP DRESSINGS

Wet wraps are a type of treatment used widely in some countries, but rather rarely in the UK or elsewhere in Europe. In recent years, we have found them an exceptionally valuable addition to the range of treatments we can offer children with extensive atopic eczema. The basic principle is that a wet dressing is applied directly to the skin, and left in place for several hours following application of a cream or ointment. There are many modifications of this general approach, and I will describe the technique used in my own patients at Great Ormond Street, which I have found practical, effective, and safe even for long-term use in children (*Figures 48a–h, 49a–d*).

Although wet wraps can be used to some effect with just a moisturizing cream, they work much better if a mild steroid is applied under the

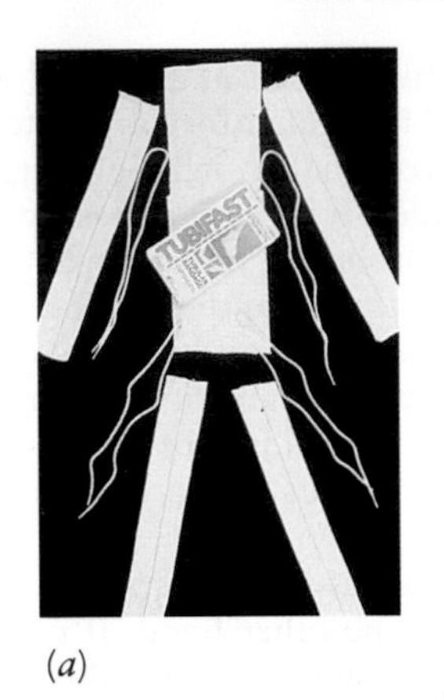

(*a*)

(*b*)

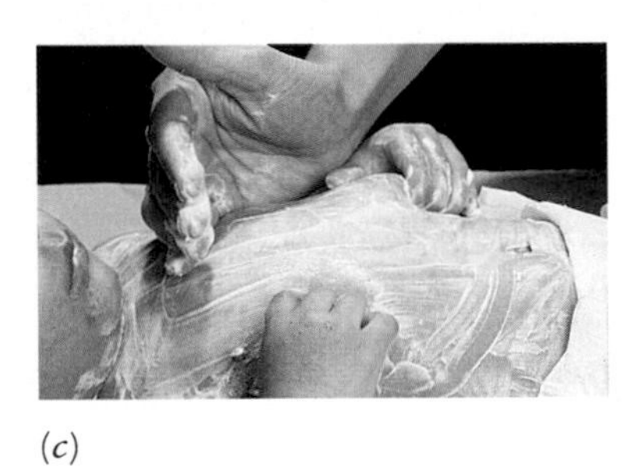

(*c*)

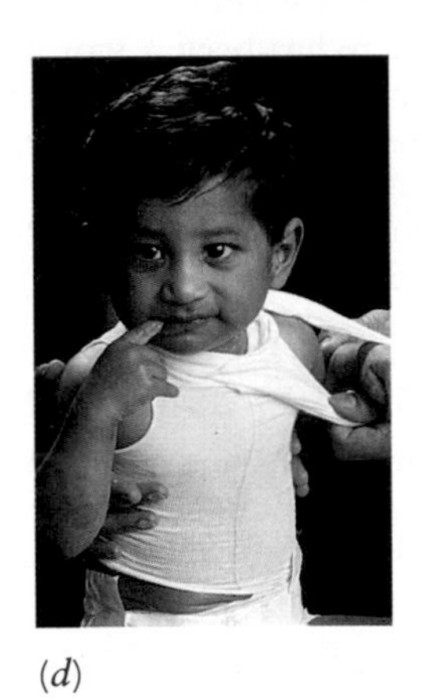

(*d*)

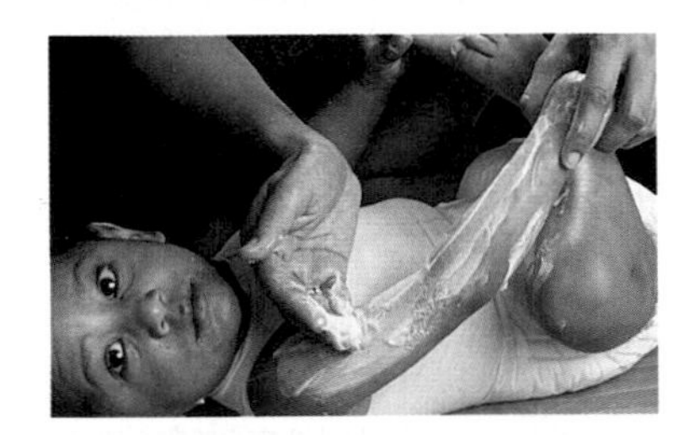

(*e*)

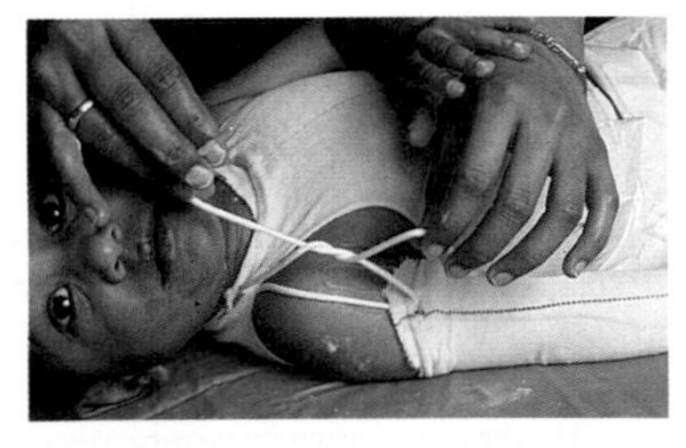

(*f*)

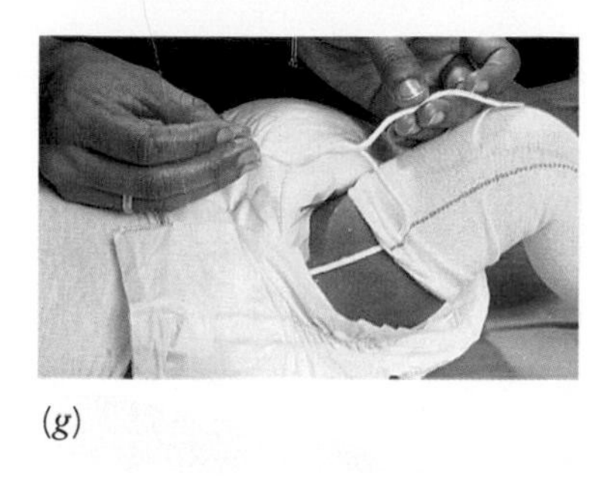

(*g*)

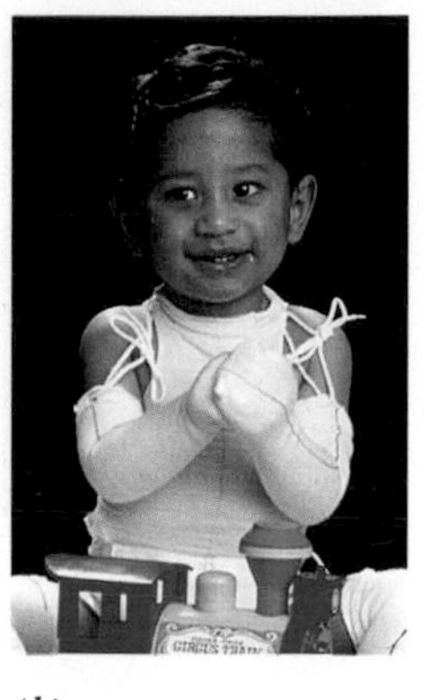

(*h*)

Figure 48 How to apply wet wrap dressings on the trunk and limbs. (*a*) Start by cutting the lengths of Tubifast® bandage you will require, two each for the trunk and the limbs. Also cut a few extra strips or use white shoe laces as ties. (*b*) Soak one set in warm water. (*c*) Deal with the trunk first. Apply the cream, generously. (*d*) Gently squeeze excess water from the wet trunk bandage and put it on like a vest. Then, as soon as possible, pull on a dry length of Tubifast®. (*e*) Now do the same for each limb in turn. (*f*) Cut little holes at the top of the arms and above the openings for the arm on the trunk bandage. Join these by passing a tie through the holes you have made, and tying together with a bow. (*g*) Do the same for the legs. (*h*) The finished job.

dressings. Hydrocortisone can be used, but I have come to favour using a diluted version of a preparation called 'Propaderm® ointment. Propaderm® contains a steroid called beclomethasone dipropionate, which has special qualities that may make it particularly safe for use in small children. Like other steroids applied to the skin, beclomethasone dipropionate may be absorbed into the blood. Steroids that have reached the blood will circulate and, if present in sufficient concentration, may have undesirable effects within the body. In the case of children, we would be particularly concerned about deceleration of growth. The steroid will continue to circulate and to have these effects until it is chemically inactivated, a process that occurs mainly in the liver. Whereas this chemical inactivation process may take several hours in the case of most steroids, it occurs within minutes in the case of beclomethasone dipropionate, so that it has little opportunity to do any harm. In order to minimize further any risk of harmful internal or

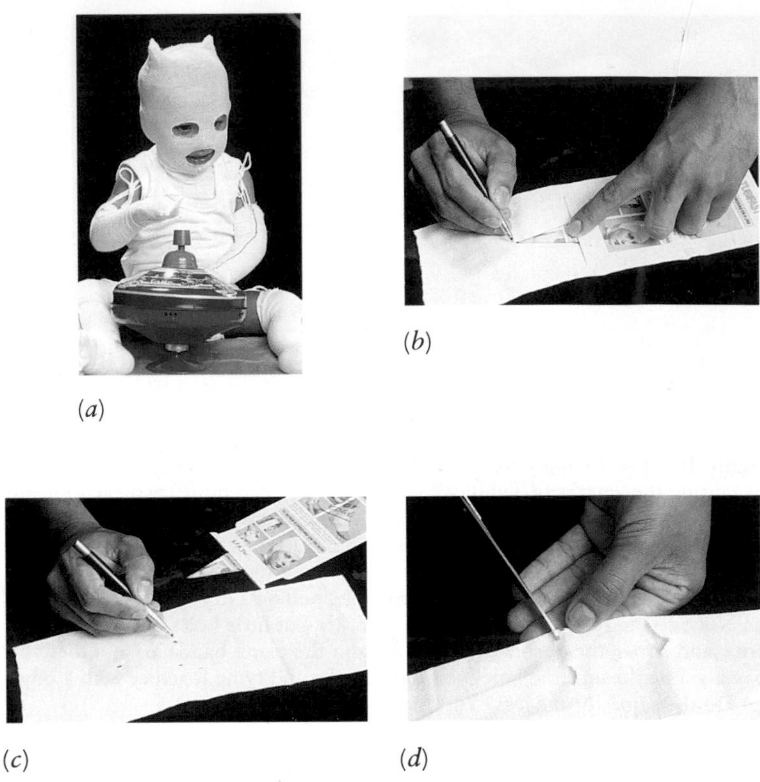

Figure 49 How to apply wet wrap dressings to the face. (*a*) If you need to use a face dressing, you will need to cut either one large, or two smaller holes, as in this child, for the mouth and nose. A neat way to finish it at the top of the head is by sewing the opening up in a line. (*b*), (*c*), (*d*) Once you have got the measurements right, you can use a piece of cardboard as a template to indicate where you should cut the holes.

external effects, the Propaderm® ointment is diluted to a tenth of its original concentration in a moisturizer called white soft paraffin BP. In many years of use, I have never seen any adverse effect with this diluted steroid ointment when used in conjunction with wet wrap dressings.

After much experimentation, *Tubifast® currently appears to be the best available tubular bandage for making the dressings. It has just the right degree of elasticity, conforming accurately to the child's contours without being tight, and retaining this elasticity well. It is also adequately absorbent, so that it can hold enough water to remain moist for several hours. Another great advantage over other similar bandages is the fact that it can be washed and reused several times before it becomes too frayed and baggy.

Tubifast® is manufactured by Seton Healthcare Group, Tubiton House, Oldham, Lancs OL1 3HS (tel. 061 652 2222), and is available in 20 metre rolls or in one metre lengths. Hospitals can provide it in the 20 metre rolls, but, unfortunately, your GP can only prescribe it in one metre lengths. Theoretically, you could buy it in either length form. Tubifast® comes in five different widths (Table 5), easily identified by a coloured strip along the length of the bandage, but, for some extraordinary and incomprehensible reason, the red-line width is not prescribable.

Before starting, you will need to prepare appropriate lengths of Tubifast® bandages, two lengths for each arm and leg, and two lengths for the trunk. You may find it helpful to measure the lengths needed against an item of your child's clothing, adding a few extra inches. It is a good idea, once you have got the lengths right, to keep one set of bandages as a template, measuring new lengths against this standard set each time.

After cutting the lengths you need, you will have sufficient to make two garments. One set should now be soaked in warm water.

This treatment should always be used after first bathing the child. It is generally best to deal with the trunk first, then each limb in turn. The diluted Propaderm® ointment is applied after a bath. In order to avoid putting hands or fingers into the pot of cream, use a clean tablespoon to take out a little more than you will require, and put this on a clean saucer. Any left over should be discarded. Instead of the usual thin application that might be used without dressings, you should in this

Table 5 Tubifast® bandages for wet wraps

Colour code	Width (cm) (unstretched)	Suitable for
Red*	3.5	Small arm
Green	5	Medium arm, small leg
Blue	7.5	Small trunk, large arm or leg
Yellow	10.75	Medium trunk, head
Beige	17.5	Large trunk

* Not prescribable by GPs

case apply the cream thickly, as if spreading butter on to bread. Generosity is essential if the treatment is to succeed. Immediately after spreading on the cream, apply the first, wet layer of tubular bandage, following equally smartly with the second dry layer. After doing the trunk, move on quickly to each arm, and then to each leg. Secure the length of Tubifast® on arms and legs to those on the trunk by using a small piece, or a white shoe-lace, as a tie. There are several alternative ways to finish the arms and legs so that the hands and feet are included. If you start with lengths that are longer than they need to be, you can twist the bandage at the finger tips or the toes, and roll the remainder back up the limb. Alternatively you can just tie knots at the ends. Some parents take lengths which are twice as long as they need and wet one half, pull this on, twist and pull back the dry half.

If you haven't put bandages on to your child before, it may be a good idea, if your child is a suitable age, to bandage a favourite doll or teddy first. You may find you have to go on doing this for a while!

These wet wrap dressings are most suitable for use overnight. Once the dressings have been completed, the best thing for the child to wear over them will usually be a one-piece poplin cotton pyjama suit, such as those made by Cotton-On Ltd (Monmouth Place, Bath BA1 2NP, tel. 0225 461155). This whole procedure can usually be accomplished in about 10–15 minutes when you become familiar with it. In the case of smaller children, the job will need two adults, at least initially. Generally, children are very comfortable in the dressings, and, often, are able to sleep well for the first time in their lives.

As you become more experienced with the wet-wrap technique, you can start to make modifications to accommodate the particular needs of your own child. You can apply a moisturizer alone, instead of the diluted Propaderm® ointment, to areas where the eczema is under good control. '50/50' (see p. 103), is ideal for this purpose, and can be prescribed separately by your GP. Have both the diluted Propaderm® and the '50/50' available as you apply the dressings, and for each area of skin, decide which would be most appropriate, like 'painting by numbers'. This has the advantage both of reducing the total amount of steroid applied to the skin, and of helping to prevent tachyphylaxis (p. 116), the tendency of the skin to become less responsive.

If a child's eczema is severe, it is usually a good idea to start off wet-wrap treatment with the dressings applied 24 hours a day, changed twice daily, morning and evening. After about a week, the eczema will be quietening down, and the dressings can be applied just at night. The night can begin whenever you wish. Most children with eczema start to become itchy at a certain time in the late afternoon, and this is the ideal time for you to put the child in the bath and then apply the wet wrap dressings. If the treatment works, the eczema will gradually settle down further, though this may take a few weeks. When you feel confident enough you could try leaving the dressings off for the odd night, and then, if the eczema allows, for a few nights at a time. Eventually, many children only require wet wraps rather occasionally, for spells when, for one reason or another, the eczema has become more active. Used in this way, the wet-wrap technique can give you the power you need to keep you in control of your child's eczema, not the other way round.

Many other modifications are feasible. In some children, the trunk area is more or less spared of eczema. In this case, you can still use wet wraps for the limbs, but you will not need to use them on the trunk. The limb bandages will however still require to be secured, and you can use either a single dry layer of Tubifast® or an old vest or T-shirt for this purpose. You can just treat the arms or legs, even just the wrists, knees, or ankles, though you will in each case need to consider how the bandage will be held in place. Micropore® tape is often helpful for this purpose, passed once around a narrow part of the limb over the outer dry layer of bandage.

Tubifast® can be washed and reused. How many times you will be able to use it depends to a great extent on how much it has previously been traumatized by your child, and it will also depend on how gently you treat it during washing. It only needs a light wash, and is best protected during washing by stuffing it into an old pillowcase or even a sock or a pair of old tights.

Many people reject out of hand the idea of bandaging up a child with eczema. They feel it must be unkind, and that it will only make matters worse by making the skin hot and therefore even more itchy. Surely, they argue, there must be a better way of treating eczema these days. This kind of prejudice is a pity because this treatment technique can undoubtedly be of great value, providing an extra dimension to treatment of children with more severe eczema. Many parents, when first told about wet-wrap treatment, do not think it sounds very promising. Nor did I, when I was first told about it by an Australian colleague. However, having tried wet wraps on a few of the children under my care, I was converted, and have been surprised time and time again by what they can achieve in many cases.

I still do not know exactly why wet wraps work so well, and can only suggest reasons. I would highlight four possible factors that may contribute to their beneficial effect:

Evaporation

The wetness is clearly important. By using the right amount of overlying bandage and clothing, the rate of evaporation of water from the wet layer can be controlled. The very gradual evaporation that results provides prolonged slight cooling of the skin, which appears to provide a high level of relief of itching.

Wetness

The wetness itself seems beneficial, independently of the evaporation. Parents of babies in nappies will often observe that their child's nappy area is more or less free of eczema, even when all the rest of the baby's skin is affected. This seems to be a result of the constant moisturizing effect of the humid conditions within the nappy. If the bandages dry out, they may become hot and uncomfortable. It is therefore important to keep them moist at all times. If they do dry out, you can take down the outer layer and wet the inner one by spraying warm water on with a hand spray, or by wetting it with a sponge. It is a good idea to check the bandages for drying out before you go to bed, and if your child awakes during the night.

Protection

Clearly, the two layers of bandage will reduce your child's access to the skin, and will therefore provide some physical protection against the harmful effects of scratching.

Enhanced steroid penetration

For the three reasons already given, the wraps work quite well with no steroid at all. However, they definitely work even better with a mild steroid. The wetness probably does help to encourage penetration of the steroid through the outer layers of the skin to the important area where the eczema reaction is taking place.

MITTS AND DRY WRAP DRESSINGS

Bandages can be used solely for their protective value. Ordinary crepe bandages may be used for this purpose, but tubular bandages are far better. A good range of tubular bandages is produced by Seton Products: Tubiton®, Tubinette®, and Tubegauz®, in addition to Tubifast®, which we have already considered. Tubegauz® is all cotton; Tubinette®

is all rayon; and Tubiton® is a mixture of the two. These are made in eight standard sizes:

- 00—small fingers and toes
- 01—larger fingers and toes
- 02—fingers and toes over other dressings, e.g. paste bandage
- 34—younger child's legs and arms
- 56—older child's legs and arms
- 78—younger child's head, baby's body
- T1—younger child's body, older child's head
- T2—older child's body.

These three bandages are thinner and less elasticated than Tubifast®; they do not wash well, and are not prescribable. For all these reasons, Tubifast® are probably preferable.

This type of bandage is most often used to make mitts, and there is hardly a child with eczema for whom this is not useful. During the day a child's hands should be free for activity, but at night mitts help to lessen the damage that can be done by scratching, without being too constraining. The technique for making mitts is illustrated in *Figure 50a–c*.

Some parents prefer to make mitts themselves from cotton material. This can be done by cutting out hand-shaped pieces of material and sewing two together around the outside. These can be made double thickness by making a second slightly larger mitt, turning this one inside out so that the seam is hidden, and putting the smaller mitt inside. If a strip of material is sewn to the outside, this can be used to attach it to pyjamas at the wrist with a safety pin. Another piece of material or ribbon can be sewn on to the wrist of the pyjama sleeve and this can then be attached to the one on the mitt by tying a bow.

Cotton mitts can be purchased from specialist cotton clothing companies such as Cotton-On Ltd (Monmouth Place, Bath BA1 2NP, tel. 0225 461155). Older children may prefer lightweight washable cotton gloves, which are available from a number of sources, including Boots Chemists, Seton Healthcare Group (Tubiton House, Oldham, Lancs OL1 3HS, tel. 061 652 2222), or from Cotton-On Ltd (Monmouth Place, Bath BA1 2NP, tel. 0225 461155).

Some parents like to use dry tubular bandages to protect more extensive areas, such as whole arms and legs. These can be secured in much the same way as already described for wet wraps. Some parents bandage their children from head to toe before they go to bed each

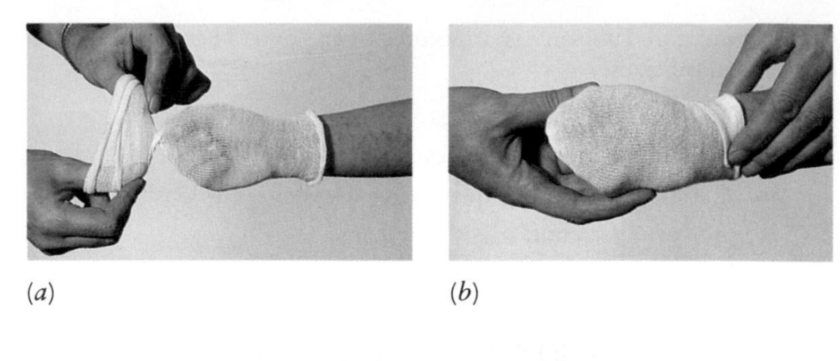

(*a*) (*b*)

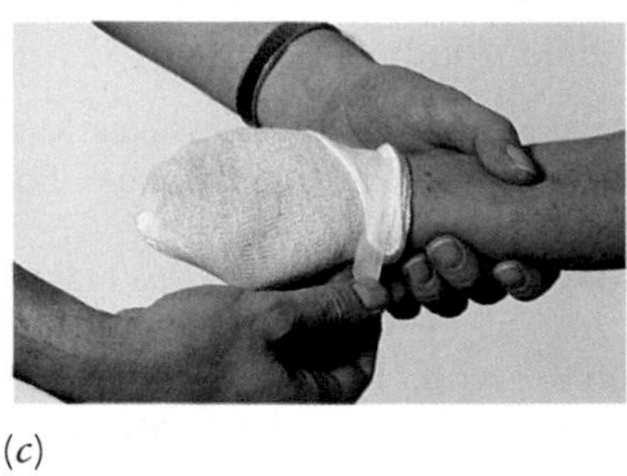

(*c*)

Figure 50 How to make mitts from tubular bandage. (*a*) Cut off an appropriate length of Tubifast® or another tubular bandage—double the distance from the tip of the thumb to the wrist. Slip on the bandage so that it reaches the wrist at one end; the same length should still be hanging free at the other end. (*b*) Twist the bandage just beyond the end of the thumbnail, with the other fingers bent. Now push back the far end over the hand. (*c*) Pinch up a fold containing both layers at the wrist. Wrap this fold around the wrist to take up the slack and tape it down using Micropore®, or another adhesive tape.

night and, though not appropriate for every child with eczema, you should consider whether protective bandaging might help your child, even in a limited form.

PASTE BANDAGES

A wide range of paste bandages is available. They are impregnated with a paste which may contain a variety of therapeutic substances. On the whole, such bandages have been manufactured primarily for the treatment of leg ulcers in adults. However, it turns out that they can also be useful for treating eczema in children, a use for which they tend to be underrated.

The most effective are those containing either coal-tar paste (*Tarband® or *Coltapaste®) or ichthammol (*Ichthopaste® or *Icthaband). They are particularly helpful when lichenification (p. 28) is prominent. Unfortunately, some children cannot tolerate the stronger and sometimes irritant coal tar, and their skin will appear to be burnt by bandages which contain it. For this reason, such bandages are usually best avoided where there are extensive raw areas.

Paste bandages which contain the fossil-fish tar, ichthammol, (p. 118) are much milder and will generally be tolerated when coal tar bandages are not. In addition, they have a more acceptable appearance and smell, and are generally preferable. *Quinaband® contain the antibiotic clioquinol, and can be helpful for infected eczema. 'Calaband® is a mild paste bandage containing calamine (p. 118), and is sometimes used in preference to tar bandages, for inflamed, angry eczema.

Paste bandages often have a refreshing, cooling effect when first applied. They can be successful in relieving irritation, and also have healing properties, particularly those containing tar. Unfortunately, they are quite messy to apply, definitely messier than the wet-wrap dressings described earlier. A second type of bandage is needed to cover and secure the wet paste bandage. Some people use ordinary crepe bandages for this purpose, but these are not ideal. I find that a rather special bandage called *Coban® (3M Health Care Ltd, 3 Morley Street, Loughborough, Leics LE11 1EP, tel. 0509 611611) is best. This is an elasticated bandage, which is self-gripping without incorporating any adhesive. It is light, allowing air to pass through easily but providing some resistance to seepage of the paste. Fortunately, it has been decided very recently that Coban® will be available, for the first time, on a GP's prescription. Some parents manage to eke out their supplies by reusing or even washing the Coban® once or twice, though after washing the elasticity tends to disappear.

Paste bandages secured by Coban® can be quite easily applied (*Figure 51a–m*). Like wet wraps, they are often best used just at night, but some children will be happier wearing them during the day as well. In this case, they can either be left on for 24 hours at a time, or when greater effectiveness is desired, they can be changed night and morning. Daytime use is possible because the Coban allows considerable freedom of movement. It seems best not to leave individual bandages on for more than 24 hours, though I have known cases where they have been successfully left on for three days at a time. They can only easily be used on the limbs, and can be used either for whole limbs or for limited parts of the limbs, such as wrists or ankles. Most children find them surprisingly comfortable. Paste bandage treatment is probably

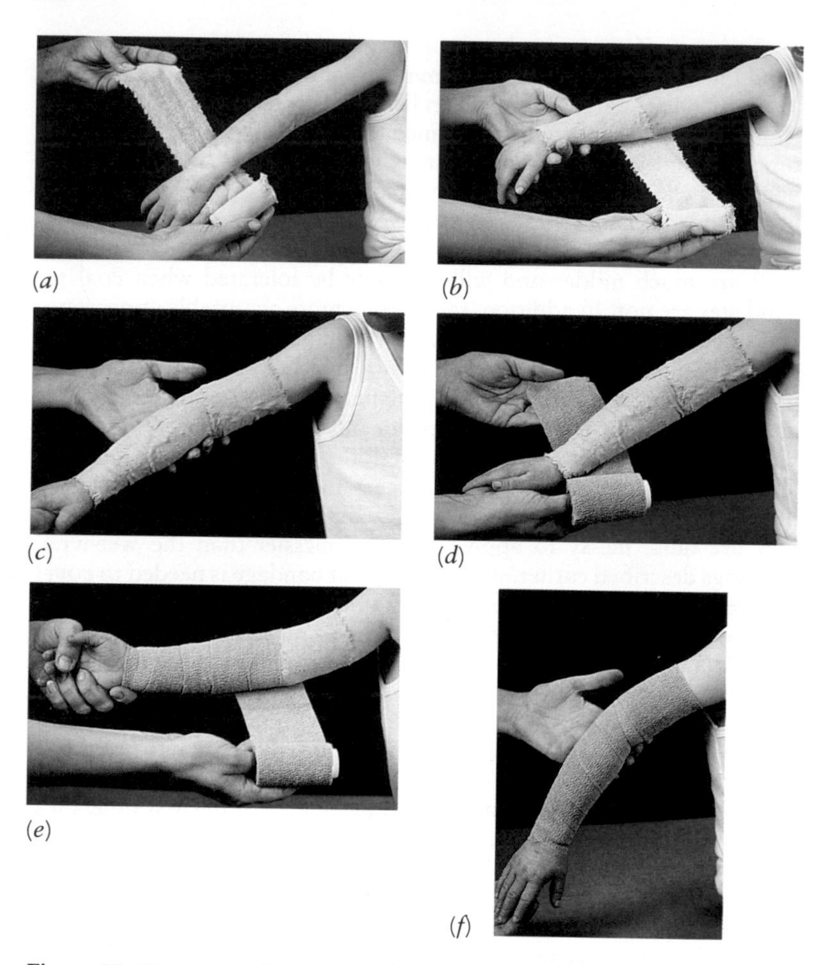

Figure 51 How to apply paste bandages. *To the arm*: (*a*) Start at the wrist. (*b*, *c*) Work up the arm. Occasionally reverse the direction of winding; this enables the bandage to slip slightly when in place, giving maximum mobility. It also helps to compensate for any shrinkage of the bandage. (*d*, *e*, *f*) Secure the paste bandage with a layer of Coban®, winding it on with a little, but not excessive, tension. Ideally, allow the paste bandage to protrude slightly at the ends.

To the leg: (*g*) Start at the foot, which is best bandaged separately. (*h*) Now bandage the ankle, again separately. Cutting the bandage avoids kinks, gives mobility at the ankle, and compensates for shrinkage by allowing the bandage to slip a little. (*i*, *j*) Now bandage up the leg, occasionally reversing the direction of the winding to allow some slipping. (*k*, *l*) Secure the paste bandage with a layer of Coban®. (*m*) The finished job!

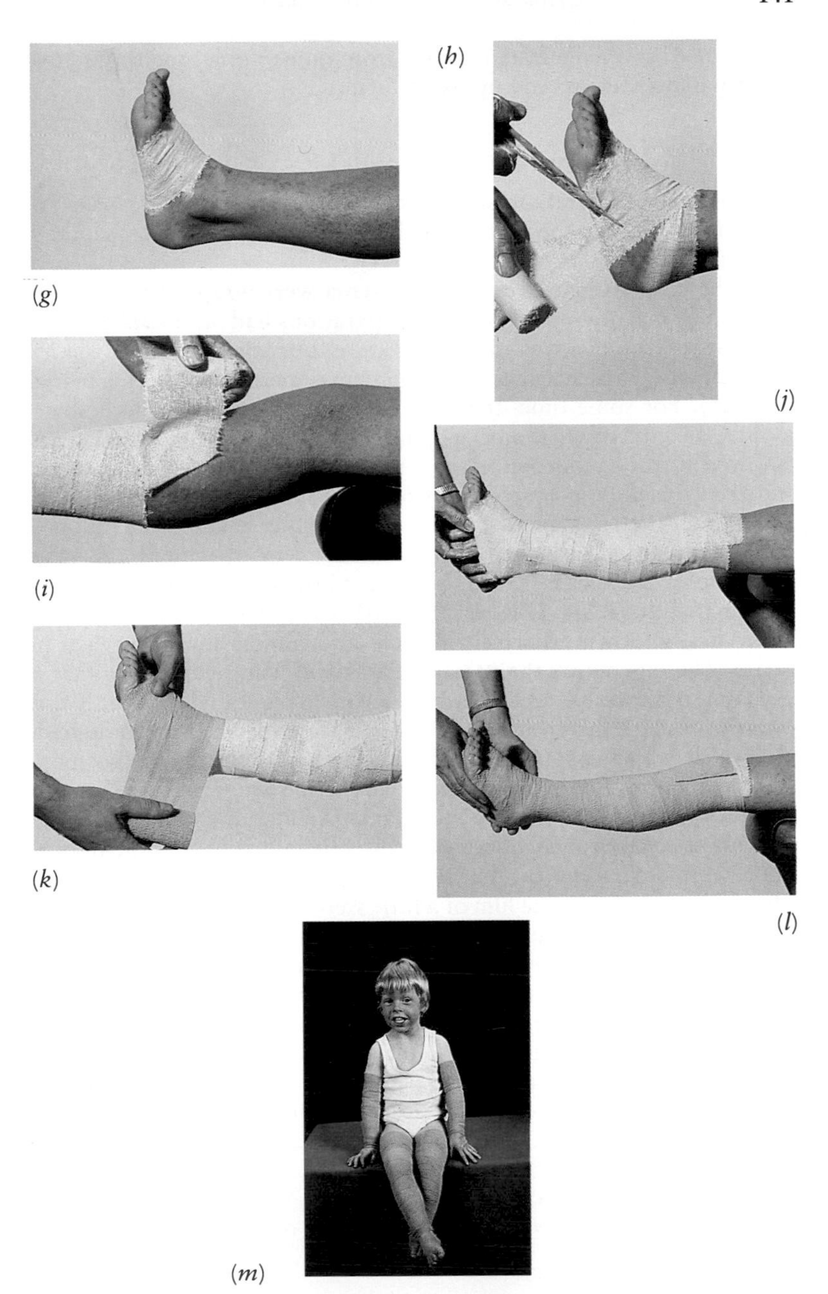
(g)
(h)
(i)
(j)
(k)
(l)
(m)

ideally used as a temporary measure from time to time, and this is how most parents seem to end up using them.

OCCLUSIVE AND SEMI-OCCLUSIVE DRESSINGS

Another use of special dressings is specifically to enhance the penetration of steroids into the part of the skin in which the eczema reaction is taking place.

In the past, patients with severe eczema were wrapped up in polythene sheeting after potent steroid preparations had been applied. This treatment technique was rapidly effective, but could result in steroid skin damage, particularly the development of permanent stretch marks (p. 114). For some time, this type of approach was therefore abandoned, but, recently, the development of new types of material has allowed it to be reintroduced in a modified form, in which it can be extremely useful for limited areas of particularly stubborn eczema.

Several different types of *semi-occlusive* films have been developed as skin dressings, principally for the treatment of wounds. A good example is *Tegaderm® (3M Health Care Ltd, 3 Morley Street, Loughborough, Leics LE11 1EP, tel. 0509 611611). This is a self-adhesive clear film which is extremely flexible and almost invisible when in place. The film allows the passage of gas and water vapour, so that it remains comfortable and does not become soggy underneath. It can remain in place for several days and is easy to remove when desired. Although water vapour can pass through the film, it does so more slowly than from normal skin, hence the term semi-occlusive. The film alone is sometimes useful for encouraging small but stubborn areas of eczema to heal, probably because it provides some protection and the more humid environment that seems to discourage eczema. However, the application under the film of a little steroid makes it a very effective tool in this type of situation. The less complete occlusion makes the technique less hazardous than was the case with polythene, but it should nevertheless not be used in this way continuously for long periods. Steroid ointments and creams are not suitable for use under this type of film because the oil they contain will interfere with its adhesion; steroid-containing eye drops such as 'Eumovate® or 'Betnesol® are ideal. Tegaderm® dressings 10 × 12 cm can be prescribed by your GP, or bought by you, and can be cut to shape if too large.

For very small areas, a tape impregnated with a steroid #Haelan Tape® can be very useful. This is a completely occlusive polythene tape,

7.5 cm wide, which can be cut to any shape. It can be used to treat very limited areas of persistent eczema, particularly where the lesions look rather like insect bites, or where they are small and coin-shaped (*nummular* or *discoid*).

TREATMENTS FOR SPECIAL AREAS

THE SCALP

The scalp is frequently affected by eczema, but is a difficult area to treat because hair gets in the way. Fortunately, scalp eczema is often very mild, and, in such cases may require no more than a suitable shampoo. Most shampoos of this type contain coal-tar solution; popular examples include *Polytar Liquid ®, *Baltar®, *Alphosyl®, *T-Gel®, and *Genisol®. However, some will find coal tar irritating, and will prefer to use just a very mild shampoo such as ^Neutrogena® or a baby shampoo. I find that an anti-yeast shampoo called 'Nizoral® (see p. 57) is often helpful for scalp eczema. These shampoos should be used at least on alternate days.

A variety of preparations are available where more intensive treatment of the scalp is necessary. In order to make treatment less messy, steroids for use on the scalp are often made up in liquid form as lotions. Unfortunately, these generally contain alcohol, which may prove irritating to those with atopic eczema, and all of them contain steroids which are in any case too potent for long-term use in children. It is important to be aware that absorption of steroid is particularly rapid through the scalp because of its vigorous blood supply. One per cent hydrocortisone lotion was the safest steroid preparation suitable for scalp use, and had the advantage of not containing alcohol, but sadly is no longer available.

For the present, therefore, treatments for scalp eczema are less than ideal. My impression is that the best approach is to work at improving the eczema elsewhere, where this is easier to achieve, and it will often be found that the scalp eczema will simultaneously settle down in sympathy. For small children, daily cleansing and moisturizing of the scalp with a water-dispersible cream is a good idea. *Diprobase® cream or *Unguentum Merck® are very suitable for this purpose. For older children, one of the shampoos mentioned above will usually be preferred. If the scalp remains dry, it can be very helpful to apply coconut oil once or twice daily. Coconut oil has a melting point very close to body temperature, which means that it is fairly solid when applied (as long as it has been kept somewhere fairly cool), and turns to liquid once on the skin. This helps to spread the ointment in the scalp. The application process is rather time-consuming. For the best results, the hair should be parted and the ointment rubbed into the scalp where it is exposed along the parting. Further partings should then be made progressively across the scalp.

THE HANDS

When one considers the stresses to which children subject their hands, it seems amazing that children with eczema do not all have their hands affected. Sadly, though, many do, and in some cases hand eczema can become a major problem. Since contact with irritants can play a major role in aggravating hand eczema, this should be minimized. Handling of fruit and other raw foods should be discouraged. The hands should be kept out of water as far as possible, except for washing as described below. Tasks such as washing up or washing the car should be forbidden. Soap is a particular taboo. One should try to direct the child's play away from certain particularly irritating activities, such as playing in sand-pits, or painting if this involves getting paint all over the hands. Contact with allergens is also an important factor in maintaining hand eczema. In this respect, playing with animals is particularly likely to be a problem, and should be very strongly discouraged.

One of the most important elements in the care of eczematous hands is frequent washing using a water-dispersible cream such as *Diprobase® cream or *Unguentum Merck®. It is best to have the cream prescribed in the form of a pump dispenser rather than a tub, as it is important not to dip the hands into it, and in order to have sufficient to encourage regular use. A basin of warm but not hot water should be run. The hands should be wetted, and then a good dollop of cream should be taken and rubbed into the hands and fingers for two or three minutes before rinsing off. This exercise should be repeated several times each day. As well as keeping the hands clean and thereby discouraging infection, this method provides a moisturizing effect without leaving the hands greasy.

A steroid ointment will be helpful, and should be used either just at night or twice daily. If the hands are a particular problem area, a stronger steroid ointment may be required than would be suitable for more general use. If a stronger ointment is being used in a child, I prefer to limit this to a once-daily application at night, under cotton mitts or gloves, to avoid transferring it to the face. It is often a good idea to have four or five steroid ointments of different potency which can be used in rotation, changing preparation every week. This rotation helps by discouraging the development of tachyphylaxis (p. 116), the process by which the eczema becomes less responsive to any particular steroid, and by allowing the use of mild as well as stronger preparations, helping to prevent the thinning of the skin that might occur if stronger preparations were used exclusively.

A wet-wrap technique (p. 129) can be very helpful, and can be adapted for use solely on the hands. *Micropore® tape is useful for securing the tubular bandages at the wrist, by passing it around the wrist over the outer dry layer of bandage.

Another technique that can be very helpful when hand eczema is severe is occlusion (p. 142), though this should only be used in older children. Care is required with this method because it does involve a hazard of skin damage by the steroids, though this is almost always of a reversible type. After the application of an appropriate steroid ointment at night, a pair of polythene disposable gloves (*Disposagloves®: large, medium, or small) is worn in bed. This increases the penetration of the steroid into the deeper parts of the skin where the eczema reaction is occurring. It also produces a moist environment which encourages healing. Generally, I only advise this in teenagers who have the motivation to keep the gloves on overnight. Some will find their hands get too hot and uncomfortable, or are too wet in the morning. However, this method can be of great value to some children, who use it from time to time, for a few nights at a time, to increase the effectiveness of their treatment. In smaller children, the same technique can occasionally be appropriate. However, since disposable polythene gloves are not available in suitable sizes, it is necessary to use small polythene freezer bags applied under a sock to hold the bag in place.

Paste bandages can also be useful, but will probably only be tolerated by younger children. They should be applied overnight (see p. 138); the technique of application is demonstrated in *Figure 52a–d*.

It is also important to apply a moisturizer fairly regularly during the day; one should aim to do so every hour. What will be suitable will depend on the age of the child. It is important to let the older child select the preparation that seems most acceptable.

THE FEET

A few children will have special problems with eczema on the feet. A small proportion of these do so because they are allergic to shoe

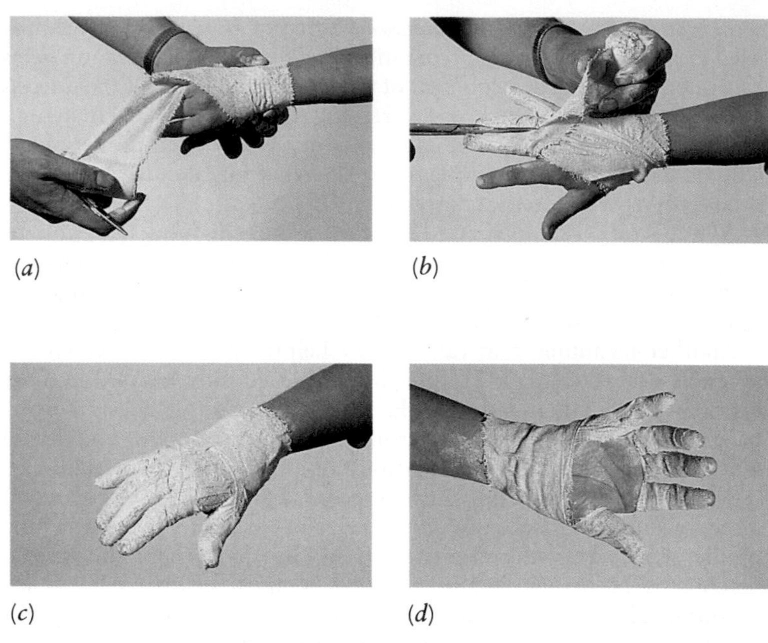

(*a*) (*b*) (*c*) (*d*)

Figure 52 How to apply paste bandages to the hand. (*a*) Start at the wrist, and then come straight across the back of the hand to the little finger. Wrap the bandage around the little finger, and then cut it. (*b*) Start again at the wrist and bandage the ring finger in exactly the same way. Then carry on to the other fingers, always starting at the wrist and cutting the bandage after wrapping up the finger. (*c*) The back of the finished hand. (*d*) The palm, which is rarely affected by eczema, will be left free, keeping the hand more comfortable.

constituents, particularly chromate, which is used in leather tanning, or the rubber chemicals used in shoe adhesives. This is the type of allergy which will need to be investigated by patch tests (p. 94). This kind of allergy is rare in children, but you definitely should be suspicious if your child's eczema affects the feet predominantly.

Fungal infections of the feet, popularly known as athlete's foot, can look very similar to foot eczema. If a child's eczema is more or less confined to the feet, parents and doctors are quite likely to mistake foot eczema for athlete's foot. The penny is likely to drop, however, when there is no response to treatment with antifungal drugs. Athlete's foot tends to start in the web space between the little toe and the next toe along. It is commoner for eczema to affect the tops of the toes, or, if a web space is affected, the space between the big toe and the next one along.

It is in practice rather unusual for children to have fungal infections of the feet, and most children with foot eczema are not allergic to shoes. Hot sweaty conditions may be one of the main reasons that a particular child gets foot eczema, and such conditions are very likely to make it worse. Footwear should be dry and permeable to air. The foot's worst enemy in this respect are shoes that are completely airtight. This is often the case with cheaper shoes, which may be made of plastic rather than leather, or may be treated with an airtight scuff-proof finish if they are made of leather. Similarly, cheaper trainer-type shoes are likely to be made of airtight plastic.

Unfortunately, the right type of shoe tends to be expensive (*Figure 53*). It should have leather uppers, or if it is a trainer-type shoe, it should be made either of leather or of fabric. If there is a lining this should also be of leather or fabric, and should be glued by the 'spot' method, not by a continuous layer of adhesive. The ideal material for the in-sock (the floor of the shoe that is in contact with the sole) is also leather. It may be difficult to check all these things without actually taking the shoe apart—not generally a popular move in shoe shops. The safest thing is to buy shoes made by a manufacturer that takes trouble to get these things right. Clarks Shoes are one of the leaders in this respect. Good leather sandals are also very suitable, though clearly they will not be appropriate throughout the year.

Figure 53 Diagrammatic section of a shoe.

If your child's shoes do not have leather in-socks, you can help to increase air circulation in the shoe by buying an untreated leather in-sock to insert over the existing in-sock. These may be difficult to find, and a cork in-sock is the next best thing so long as it does not make the shoe too tight. Many shops also sell special 'ventilators' which do the same job.

If the same shoes are worn day after day, there will be a steady build-up of moisture in the leather. This should be avoided by giving shoes a day's rest for every day they are worn. This also makes for additional expense, but does have the advantage of extending the life of the shoes.

Socks act as a wick, drawing moisture up from the foot to the ankle and lower leg, from where it evaporates away. Long socks do this better than short ones, and those made of natural fibres do it better than artificial ones. An alternative is good quality towelling sports socks.

THE NAPPY AREA

Many normal babies get rashes in the nappy area. However, although atopic eczema does sometimes start in the nappy area, more characteristically other areas are affected, and the nappy area is relatively spared. One of the main reasons for the sparing of the nappy area may be the high humidity, and the presence in the urine of natural moisturizing substances, particularly urea (p. 103). If, on the other hand, your baby is having problems in the nappy area in addition to typical atopic eczema elsewhere, the odds are that the problem is that your child's urine is for some reason proving unusually irritant. An important way to tackle the problem is to change the nappies more frequently, to try to make sure that wet or soiled nappies remain in contact with the skin as briefly as possible. You must make sure to wash well whenever the skin is soiled, using a dispersible cream such as *Diprobase® cream or *Unguentum Merck®, as described above for hand eczema. In addition, at every nappy change you should apply a generous layer of a suitable protective moisturizer such as white soft paraffin (Vaseline®) or a mixture in equal parts of white soft paraffin and liquid paraffin. If these measures are not effective, there may be a problem with yeast infection, which can complicate eczema in the humid conditions of the nappy area; in this case, your doctor would need to prescribe an ointment combining a topical steroid and an anti-yeast agent.

THE GENITAL AREA

The genital area is frequently affected by eczema, in both boys and girls. Parents often worry that their child will do serious damage by tearing at the scrotum or labia, and they may also find that their child's interest in the genital area is embarrassing, especially when the scratching takes place in front of other people. In fact, real damage is never done. The genital area tends to be a slightly itchy area in all children, and this itchiness is merely exaggerated in those who also have eczema. Eczema may become established in this area, but can be treated in the same way as eczema elsewhere. Parents worry that the genital area may be particularly liable to harmful effects if topical steroids are applied, but this is not in fact the case. In reality the genital area seems to be tough, and no unusual precautions need to be taken.

ANTIHISTAMINES AND THE SLEEP PROBLEM

The itch of eczema results from the release of chemical substances into the skin during the inflammatory reaction (see p. 11). These chemical substances are made within the various types of cells involved, and possibly also within the nerves that supply the skin, and are often known technically as *mediators* of inflammation. Unfortunately, it is not clear exactly which of the many mediators that are released is actually responsible, and it is possible that there are several. At one time it was thought likely that the principal cause of itching is a mediator called *histamine*. Drugs which could counter the effects of histamine (*antihistamines*) therefore became a popular treatment for eczema, and they did appear to have some modifying effect on the itchiness of the condition. However, it has gradually become apparent that their effects are almost entirely a reflection of the sleepiness (*sedation*) they induce rather than any effect on the skin itself. This lack of direct effect in the skin suggests that mediators other than histamine are probably more important.

There are now a variety of antihistamines which cause little if any sleepiness (*non-sedative antihistamines*), which are very useful in certain medical conditions because high doses can be given without the risk of drowsiness. However, because it is this sedative effect which is itself responsible for the benefit, albeit limited, of antihistamines in atopic eczema, these non-sedative varieties generally appear to have no beneficial effect.

In adults, the sedative antihistamines almost invariably cause appreciable drowsiness, and they are in fact quite often used to treat insomnia. This effect is rather less predictable in children, and some will instead experience a stimulating effect, which will make them more excitable during the day and sleepless at night. Nevertheless, most children do experience the drowsiness, which is likely to cause problems during the day, such as irritability, lack of attention at school, and a slowing of reactions which can be potentially dangerous (when cycling or crossing roads, for example). For this reason, antihistamines are probably best avoided during the day except in special circumstances, for example, when a child is in bed or in hospital. On the other hand, the sedative effect may be turned to advantage when antihistamines are used at night, though it may be rather prolonged and persist into the following morning. The night-time use of sedative antihistamines should be considered for any child with eczema who has difficulty getting to sleep, or who wakes regularly during the night. They should also be considered for children who seem to get to sleep without difficulty and who do not often wake fully during the night, but who sleep so lightly that they do a great deal of scratching while they appear to be asleep.

A wide variety of sedative antihistamines are available, among the most frequently prescribed being 'Vallergan®, *Phenergan®, *Piriton®, or 'Atarax®. Often relatively high doses are required to help eczematous children, and one of the commonest reasons for them not working is that the dose given is too small. They should generally only be given at night, ideally about an hour before your child goes to bed. This will give time for the sedative effect to appear, and the earlier you give it, the less likely a 'hangover' the next morning will be. The dose should be increased or decreased to get the optimum balance between a good sedative effect overnight and undesirable drowsiness in the morning. A morning 'hangover' can often also be avoided by giving it still earlier in the evening.

Antihistamines are not addictive and, when they are no longer needed, their use can gradually be discontinued without problems, but it is often best to do this over a period. They are safe and have been used for many years in thousands of children without any significant long-term hazard coming to light.

However, it is often found that, with regular use, an antihistamine may appear to become less and less effective—the body seems to get used to it. This does not happen in every child. Up to a point, it can be overcome by an increase in dose but, in some cases, the antihistamine eventually seems to have little effect at any dose; this is a sign that the drug should be stopped for a period. After an interval, the effect will usually be regained. Bearing this problem in mind, a case can be made for using these drugs intermittently, either just on particular nights when you think your child may have difficulty sleeping, or during phases when the eczema is especially troublesome. If you are giving antihistamines every day, you will find them less effective when they are most needed.

Some children just don't seem to get any benefit at all from antihistamines and, as I have already mentioned, some even become rather hyperactive under their influence. In these cases they are best avoided. Unfortunately, other sedative drugs don't seem to be as helpful in eczema, and children who get no benefit from antihistamines are rarely helped by these other drugs.

CHINESE HERBAL TREATMENT

In the last few years, it has become clear that atopic eczema can, in many cases, be treated very effectively by doctors trained in traditional Chinese medicine. As a result, research has been started on this form of treatment in my hospital, at the Royal Free Hospital, and at the Middlesex Hospital. Interest in these medicines arises from the pressing need for more effective and safer treatments for children and adults with atopic eczema which has not responded adequately to orthodox treatment.

The Chinese have a sophisticated medical tradition that preceded the development of anything similar in Europe by over a thousand years. During these hundreds of years, the Chinese developed an extensive range of treatments based largely on native plants. These evolved very gradually, and considerable knowledge of the benefits and hazards of the various parts of all these plants has now been accumulated.

China probably has the richest variety of medicinal plants of any country in the world, and the most highly developed expertise in the use of such materials. However, there is a real danger that, in its keenness to emerge into the late 20th century, China might lose a tradition that has little scientific basis and which may be perceived as making it appear backward. If this were to happen, knowledge of the cultivation and use of these medicinal plants might be lost.

Herbs, minerals, and some animal products are used in China to treat a wide range of human diseases. The aim of treatment is to restore harmony to the functions of the body. The choice of treatments is determined by the effects each one is believed to have on body function. Generally an individual patient's treatment will consist of a mixture of several herbs which are given simultaneously. Although a limited number of formulations of these agents are available in pill, tablet, or liquid form, mostly they are prescribed in dried form. The treatment is prepared by boiling these herbs with water for a specified period, straining off the liquid, and drinking this after it has cooled.

Although the theory and language of traditional Chinese medicine appears antiquated to doctors and scientists in the West, the question of possible beneficial effects of the treatments used in certain diseases requires very serious consideration. The possible therapeutic effectiveness of this type of treatment was first brought to my own attention several years ago, when a boy under my care with severe atopic eczema was taken to see a Chinese doctor in London's Chinatown area. He was treated with a daily decoction prepared from a mixture of dried herbs. This treatment proved impressively successful for him, and there were no obvious adverse effects.

Subsequently, more than 30 other children under my care were treated in this way by the same Chinese doctor, and we observed sustained and marked improvement in about 75 per cent of them. In many cases, the eczema was severe and had not been adequately controlled by any of the full range of currently available orthodox treatments in the UK. Other dermatologists who care for children and adults with atopic eczema observed the same improvements in their own patients who were treated by this and other Chinese doctors trained in this form of treatment.

It became clear that treatment with decoctions of Chinese medicinal herbs might be generally more effective than the conventional treatments used for atopic eczema in Europe and North America. Furthermore, careful observation of patients treated in this way failed to reveal evidence of any obvious toxicity. I therefore approached this Chinese doctor, Dr Ding-Hui Luo, who was trained in traditional Chinese medicine in Canton, and she responded by offering her time and expertise without hesitation. She has given extremely valuable advice over the last few years. In common with other doctors of traditional Chinese medicine, she generally prescribes individual formulations for her patients, each containing 8–10 different herbs, which are chosen from a wide range (*Figure 54*).

At my request, Dr Luo agreed to devise a limited number of standardized formulas for patients whose eczema and general condition fulfilled certain fairly precise criteria. She prepared five formulas for mixtures of medicinal herbs that we could use in this way, and she also provided descriptions of the type of patient for which each of these five formulas would be most appropriately prescribed. Her husband, Mr Lau, is a Chinese pharmacist, and he was able to provide both a Latin name for each component and the appropriate dosage for patients of different ages.

We have been able to verify his identifications with the help of the laboratories of the Royal Botanic Garden and botanic experts in China, and we now have the precise species names for most of the plant materials we use.

Figure 54 A day's Chinese herbal prescription.

The next stage was to prescribe these formulas to patients whose eczema had not responded adequately to conventional therapy. The prescriptions were taken to Mr Lau, who dispensed them. These pilot studies suggested that these standardized formulas would be helpful in as many as 60 per cent of correctly selected patients. Furthermore, blood tests on these patients demonstrated no evidence of adverse effects on the liver, kidneys, or bone marrow function after six months of continuous administration.

We felt we were now in a position to undertake a full-scale clinical trial of one of these formulas, and we were determined to make the trial *placebo-controlled* and *double-blind*, so that the results would be accepted by the Western scientific community. Placebo-controlled means that the treatment under trial is compared with a dummy (or *placebo*) treatment that is as nearly identical as possible in appearance, taste, and smell. Double-blind means that neither the doctors conducting the trial, nor the patients or their parents, know at any time which treatment they are receiving, the real treatment or the dummy treatment.

The design of a trial of this treatment approach, in which prescriptions normally differ considerably from patient to patient, was very challenging. Ethical considerations dictated that the trial should be undertaken in patients whose disease had not responded to conventional treatments. We decided to limit participation in the study to children with widespread and severe eczema which had failed to respond adequately to conventional therapy. The children we invited to join the study had a more or less uniform presentation of eczema, in which the disease was very extensive, dry, and not weepy or infected. Our plan was to employ one of the standardized formulas provided by Dr Luo, with every child receiving identical treatment, despite our awareness that this method would differ from the normal practice of traditional Chinese medicine.

One of our great problems was to finance the purchase of the medicinal herbs from China, and to achieve some degree of quality control. It seemed to me that this was a task for the pharmaceutical industry. I approached an English doctor whom I had known for some time, who had recently set up his own pharmaceutical company. In the past, he had worked for a French firm specializing in natural plant medicines, and I knew he would understand and sympathize with our problem. I was delighted when he almost immediately offered to help us himself, in exactly the way we wanted, by buying, checking, and packing the plant materials needed for our trials. After a few months, he set up a small new company called Phytopharm (*phytos* means 'plant' in Greek) to support us, and recruited a pharmacist with a great deal of experience of medicinal herbs to head it.

The real (or 'active') treatment contained a mixture of parts of ten different plants, all in common use as medicines in China. These were checked for quality in various ways, for example by checking them for any content of poisonous metals such as mercury and arsenic, which might have been present as accidental contaminants or might have been added intentionally, as they are still occasionally used in various parts of the world as treatments for skin disease. They were then prepared by milling to a fine powder, carefully mixing in the right proportions, and sealing fixed amounts of the final mixture in porous paper sachets resembling large teabags. Two types of sachets were prepared, larger sachets containing the majority of the constituents, and smaller ones containing those which incorporated volatile components that would otherwise be lost during prolonged boiling.

The placebo comprised a mixture of inert plant materials having a similar appearance, taste, and smell, but with no known benefit in atopic eczema. These included bran, hops, barley, and small amounts of culinary herbs. As you can imagine, we had some fun sessions tasting and modifying the placebo mixture until we could not distinguish it from the active mixture. This was when I learned just how distasteful these mixtures actually are!

We used a trial design known as 'cross-over', which means that every child would receive both the real ('active') treatment, and the placebo treatment, each for eight successive weeks, with an intervening four-week period in which the effect of the first treatment would be 'washed out'. The order of the treatments was decided in a random way, that was only known to a third party, not to the doctors involved in the trial nor to the children or their parents. The children attended the clinic ever four weeks throughout the five-month period.

We were able to persuade 47 children of both sexes to take part in the study. Before treatment started, all the children had blood checks to make sure that their bone marrow, kidneys, and liver were functioning normally.

The parents were instructed to prepare the medication once daily, the number of sachets depending on the child's age. The large sachets were placed in about half a litre (a pint) of water which was brought to the boil and then simmered for 90 minutes. The small sachets were then added and the liquid simmered for a final three minutes. Parents learned to boil the liquid at a rate that would reliably reduce its volume to about half a glassful, and this was taken by the child while still warm.

At each visit, the children's skin was assessed using a simple scoring system, based on one we have used in previous studies. At the end of each of the two treatment periods, another blood test was taken, and a 24-hour collection of the child's urine.

Of the 47 children who started the trial, 37 were able to finish it. Five were not able to take the treatment as instructed because its taste was so dreadful, and five others had not been able to complete the study for other reasons, such as having to be admitted to hospital for attacks of asthma or skin infections.

At the end of the study we handed the results over to an independent statistician, who had a copy of the secret codes which recorded the order in which each child had taken the two treatments. Analysis of the results showed that the active treatment was much more effective than the placebo, and that it was in fact a very successful treatment for most of the children. We regard a 60 per cent improvement in a child's scores as indicating a very worthwhile benefit from treatment, and it was calculated an improvement at least as good as this was observed in the majority of the children. We were also able to see that most of the improvement was already apparent by the fourth week of treatment. The improvements we recorded in scores were matched by improvements reported to us by parents during the study in terms of the children's ability to sleep, for example. The treatment appeared to have had no effect on asthma, either beneficial or otherwise.

We were pleased to find that no child had developed any abnormality of bone marrow, kidney, or liver function during treatment.

Almost identical benefits and lack of side-effects have been seen in a similar trial in adults at the Royal Free Hospital in London. The results of these trials have verified the effectiveness of Chinese medicinal herbs as a treatment for atopic eczema. This is extraordinary for two reasons. First, it is extraordinary because it means we have the first effective new treatment for atopic eczema since topical steroids were introduced some 40 years ago, other than a small number of powerful drugs whose use is associated with serious toxic side-effects. Secondly, it was exciting because the eczema we treated had previously proved impossible to control adequately with the orthodox treatments used in dermatology clinics throughout the West.

It should however be emphasized that the patients who took part in the trials were carefully selected. They all had eczema that was widespread and severe, but free from weeping or obvious infection. It is not yet clear whether this standard herbal formula would be as effective in all forms of atopic eczema. Furthermore, about a third of patients involved in the trials did not experience a real improvement.

The herbal brew tastes terrible, and few children under the age of four years are prepared to take it voluntarily. However, research is now in progress to develop and test a new version of the treatment in which it is given as tiny pills.

No-one yet knows exactly how this treatment works. It is likely that it contains a number of active substances, and research is already under way to try to identify these. We were very keen to make sure that the herbs did not work merely because they contained some kind of steroid, and tests undertaken on the urine of all the children that took part in the study have shown conclusively that this is not the case. An enormous amount of work is still needed, both with patients and in the laboratory, where it is going to be a long and arduous task to try to identify the substances in these plants that are providing the benefits we have seen.

It is important to be aware that being a 'natural' treatment does not in itself provide any real guarantee of safety. Chinese doctors have been using the plants in our formula for many centuries, and inform us that they are very safe. We have seen very little evidence of any harmful effects in our trials. Nevertheless, the anxiety remains that a small percentage of individuals might develop some kind of undesired effect from treatment in the longer term. This anxiety has been heightened by a few letters to medical journals suggesting that hepatitis has occurred in patients taking Chinese herbal treatment for atopic eczema. Whether or not the hepatitis in these patients was due to the herbal treatment is not clear. Any patient considering Chinese herbal treatment must be aware of the possibility of harmful effects, even though there still is no definite evidence that such effects ever occur in practice. Careful monitoring during treatment is therefore advisable. Furthermore, because of these doubts, it seems appropriate, for the time being, only to recommend this form of treatment to those individuals for whose eczema conventional treatments have failed to provide adequate benefit.

Despite these unresolved issues, we have seen that Chinese herbal treatment offers a chance of substantial benefit to patients with troublesome eczema before progressing to other treatments that are associated with greater risks.

HOW TO GET CHINESE HERBAL TREATMENT FOR ECZEMA

No form of Chinese herbal treatment is yet licensed for prescription in the UK or any other Western country.

Chinese herbal treatment can be obtained through the small number of Chinese doctors trained in traditional Chinese medicine who work in the UK. If a patient is contemplating a consultation with such a doctor they should check that the practitioner has the proper qualifications, though this is not easy. A few doctors who are skilled in the treatment of atopic eczema with Chinese herbal treatment, and whose qualifications are known to me, are listed below. This type of treatment will be quite expensive, currently at least £3 per day. Patients may be at increased risk of a harmful reaction if tests of liver and kidney function are not carried out before treatment is started, and at regular intervals during treatment, but it appears that such reactions are probably rare if the doctor is properly trained and the patient is in good health.

It is also theoretically possible for hospital specialists to prescribe the specific mixture of herbs that was used in the trials mentioned above, if they consider that this would be in a patient's best interests. This treatment has the additional advantage that the quality and purity of the herbs has been checked, giving an extra margin of safety. We would recommend that a patient's suitability for this formulation be evaluated by a dermatologist. This treatment is manufactured and supplied by Phytopharm Ltd, 16 Eastbourne Road, Hornsea, East Yorkshire HU18 1QS (tel. 0964 536402), under the brand name +Zemaphyte®. This company will supply Zemaphyte® to hospital pharmacies.

Doctors are also able to write private prescriptions for Zemaphyte®, though the cost to the patient will then include pharmacy charges.

Eventually, I would like to see this treatment available more widely to children and adults with eczema, but this will probably have to wait until it is licensed by the Department of Health in this country, and by equivalent bodies in other countries. Because it is a very different kind of treatment, obtaining such licences may require changes in the licensing rules, and this means that this is all likely to take some time. Nevertheless, during the lifetime of this book, Zemaphyte®, or its minitablet successor, may be granted a product licence allowing its prescription more freely. If you want to check on the situation, please write to Miss Jayne Allan at Phytopharm at the address above.

WHICH CHILDREN SHOULD NOT TAKE THIS FORM OF TREATMENT?

Chinese herbal treatments should not be taken by children with known liver or kidney disease, or by children taking any regular medication by mouth.

THE FUTURE

We believe that there is an important role for Chinese herbal treatment, because it may provide benefits for patients with certain diseases for which conventional Western treatment is unsatisfactory. There will be a role for ready-made mixtures of herbs in certain disorders, and a role for doctors who are trained in the prescription of herbs on a more individual basis. There will be a need for controls on the quality of doctors who offer Chinese herbal treatment, and controls on the quality of the herbs themselves.

There is a need for more collaboration between doctors and scientists in the West and experts on Chinese herbal treatment, and a need for a great deal more research to bring its potential benefits to many more patients with a wide variety of diseases. This research costs a great deal.

TRADITIONAL CHINESE MEDICINE CLINICS

I am aware of the following properly trained practitioners who are qualified to prescribe Chinese herbal medicines:

Dr Ding-Hui Luo, 15 Little Newport Street, London WC2, tel. 071 437 4910

Dr Guang Xu, Crysanthemum Clinic, 3 Station Parade, Burlington Lane, Chiswick, London W4, tel. 081 995 1355
CHI Clinic, Riverbank House, Putney Bridge Approach, London SW6, tel. 071 371 9717. This clinic employs Chinese doctors who change from time to time but are all properly qualified.

TREATMENT OF UNDERLYING ALLERGIES

Clearly, the most logical and satisfying way of treating a child's eczema would be to identify the causes, and then eliminate them. Unfortunately, this isn't possible for the present. We remain uncertain to what extent atopic eczema is caused by allergic reactions and, in the absence of appropriate and reliable allergy tests, we are unable to identify the responsible antigens. We do not understand the interactions between these allergic reactions and other factors that we know also play a role in causing the disease. It is also clear that complete avoidance of the allergens which are most likely to be of greatest importance would in many cases simply not be feasible.

The difficulties in allergy diagnosis and treatment are such that I have come to the view that it is generally best to try to control atopic eczema by other means. Only when this approach fails are the difficulties of attempting to sort out underlying allergies really justifiable. However, having said this, the effort is sometimes worthwhile, and can be exceptionally rewarding.

As we have already discussed in Chapter 5, the allergens that provoke atopic eczema tend not to be the more exotic things like lobster or penicillin; they are almost always very ordinary things that are familiar components of our daily lives. They are everyday foods like eggs and milk, and universal environmental components like house dust mites, pollen, and pet dander. To make matters even more complicated, it is not usually a question of allergy to just one of these things but to several and often even most of them.

So how does one start to unravel all this? We have already considered the value of allergy tests in Chapter 5, and I will return to the subject again later. On skin testing, children with eczema generally react to a variety of common substances of the kind I have just indicated. Unfortunately, so also do some other children who do not have eczema or any other medical problem (maybe as many as a third of all children), though many of these are at risk of developing eczema, asthma, or hayfever later on. These skin tests certainly do give clues, but clues that need careful interpretation if they are to be of any practical value. What is clear is that a blind belief that the results of skin testing can precisely identify the causes of a child's eczema will almost always lead to failure in treatment. The same applies to the IgE RAST and all the other widely available allergy tests. One needs to seek other evidence that particular allergens are provoking a child's eczema. In practice, the best way to do this is to study the effect of avoidance, and to confirm any apparent benefit from avoidance by observing the consequences of reintroduction. A snag is that avoidance may not result in obvious improvement in a child's eczema unless most of the principal offending allergens are eliminated at once. This situation is somewhat analogous to that of a combination lock. Getting one number right is simply not good enough. In eczema, however, one can often achieve a good deal without identifying every relevant allergen; a good proportion will usually do.

FOOD ALLERGY AND ELIMINATION DIETS

A good place to start is with foods, since a child's food intake can be manipulated relatively easily. Remember that in practice dietary treatment is rather rarely effective as a form of treatment in children older than about four years. However, in the very young, food allergies do sometimes seem to be a major cause of atopic eczema, and avoidance diets are occasionally dramatically beneficial. Therefore, if your child has eczema which remains severe in spite of an adequate basic treatment regime, you could consider a trial of dietary treatment. However, before embarking on this, you should ask yourself again whether such treatment is justified, by considering carefully the following points:

1. Is your child's eczema sufficiently troublesome to warrant such treatment, bearing in mind the fact that it will mean forbidding foods which your child enjoys, as well as foods which are of great nutritional value?
2. Is your child of an appropriate age to give dietary treatment a real chance of success? Research has shown that it works best in the first few years of life, particularly under the age of five years, and even more particularly in the first year. In older children and in adults, the chance of a worthwhile response is much less, though a trial may still be justified if the patient is determined to have a go.
3. Have you really given simple skin treatments and avoidance of irritants a fair trial? You might also reasonably ask yourself whether it is logical to try to assess response to dietary eliminations if your child is still repeatedly exposed to a feathered or hairy pet.

4. Have you discussed the issue with one of the doctors involved in your child's care, to make sure that there is nothing else you should be trying first, and in order to be referred to a dietitian for detailed dietary advice?
5. Are you yourself really able to undertake this type of manoeuvre? You should not underestimate the additional effort involved, and, if you are already overburdened, you may find it intolerable. You will find dietary treatment of your child particularly difficult if you have a large family, or if you are already preparing different meals for other family members, such as a baby.

If you are happy on all these points, an exclusion diet will be worth considering. The main problem one has in identifying foods that may be provoking a child's eczema is that virtually any food may be involved. Questions like: 'Do tomatoes cause eczema?' are inappropriate. The right question is, 'Are tomatoes one of the things that cause *my* child's eczema?'. Some foods do appear particularly likely to aggravate eczema, but almost any food can do so. The important thing is to identify those foods that are responsible in the individual case.

Milk- and egg-free diets

Although almost any food appears able to provoke eczema, two stand out from the crowd because of their particular propensity to do so; they are eggs and milk. The only really reliable way to exclude the possibility that eggs and milk are aggravating a child's eczema is by a trial period of complete avoidance. Such a diet is an example of an *empirical* exclusion diet, which means that the foods to be eliminated have been chosen on the basis that they are statistically the most likely offenders, not because there is any specific reason to believe that they are responsible for provocation of eczema in your child's case.

Because the provocation of atopic eczema by foods may not be noticed under normal circumstances, a trial of dietary treatment may be worthwhile even if you have not noticed any aggravation of the eczema when your child eats milk or egg. Our own experience suggests that a simple empirical elimination diet may help up to 40 per cent of babies under the age of one year, up to 30 per cent of children between the ages of one to four years, up to 20 per cent of children between the ages of five to eight years, but much less often in older children and adults. When contemplating such a diet, you must remember that these certain foods provide important nutritional elements in a child's diet, especially milk and in the case of small children. The elimination of these foods could result in serious nutritional deficiency if proper compensatory adjustments were not made.

To undertake dietary treatment successfully you need the help of both a doctor and a dietitian. The doctor may be a paediatrician, a dermatologist, a clinical immunologist, or your GP. The doctor's roles are:

(1) to make sure that the diagnosis of eczema is correct;
(2) to advise on the many kinds of treatments available other than diets;
(3) to recommend dietary treatment where appropriate, and then to put you in touch with a dietitian;
(4) to prescribe nutritional supplements when these are necessary, e.g. milk substitutes, calcium tablets, and other minerals and vitamins.

In the case of children, elimination diets should *never* be attempted without the advice and supervision of an appropriately trained and experienced dietitian. The dietitian can help by:

(1) checking the nutritional adequacy of an elimination diet and to advise on any supplements that may be needed (e.g. calcium, vitamins, extra energy-rich foods, milk substitutes, etc.);
(2) providing lists of both suitable and unsuitable foods for your child's diet;
(3) supplying you with recipes and ideas for nutritious meals;
(4) helping to make the diet interesting and palatable;
(5) helping you to understand and interpret food labels, and to have up-to-date information from food manufacturers;
(6) providing encouragement and support—necessary for most parents whose children are on special diets.

Usually, it will be recommended that certain other items are excluded in addition to eggs and milk, especially chicken and selected food additives. Chicken elimination is often advised because chicken contains some proteins that are also present in egg. Avoidance of tartrazine and similar colourings, plus some preservatives such as benzoate, is also frequently recommended, as these appear able to aggravate eczema in some cases.

If such a diet is going to help, it will usually do so within six weeks.

The full benefit may not be seen if the diet is abandoned too early. If it proves helpful, you should try to identify the foods able to aggravate your child's eczema by reintroducing the eliminated items one by one. It may take several days to appreciate any deterioration in the eczema when a food is eaten which is playing a role in causing the eczema, and the food may need to be eaten on several successive days in order to provoke a detectable deterioration. For these reasons, a full week should be set aside for attempted reintroduction of each item. With milk, for example, one teaspoonful (5 ml) only should be given on the first day. If no adverse effects are apparent following this first dose, give between half a pint (250 ml) and one pint (500 ml) on each of the next six days. If at any time during the week you feel that the eczema is becoming worse, stop the reintroduction at once. If you feel that there may have been an unrelated explanation for an adverse reaction, you could try again later. At the end of the week, if there has been no change in the eczema, you should not worry about reintroducing milk into the diet on a more permanent basis. It is entirely wrong to expect to be able to tell whether your child's eczema is aggravated by a food if you only give a small amount from time to time. Foods often have to be given in quite large amounts on successive days to 'unmask' the allergy.

Every year you should attempt again to reintroduce any foods you are still eliminating, as you will often find that by then the allergic response has disappeared.

It is sometimes justified to undertake a trial of avoidance of certain foods that seem able to aggravate atopic eczema rather more often than would be expected by chance. These foods include fish, wheat (not just gluten), and citrus fruits.

Implications of eliminating cow's milk

Cow's milk is a very important element in the nutrition of all children, especially the younger ones. Its importance lies in its richness in two things; protein and calcium (vital for normal growth of bones and teeth). In order to guarantee adequate amounts of these nutrients on an egg- and milk-free diet, alternative sources will need to be present in the diet. Alternative sources of protein include meat, fish, pulses (beans and lentils), and complete milk substitutes. Calcium can be provided in the form of a complete milk substitute, or in tablet form. Currently there are three categories of complete milk substitute: soya 'milk', casein or whey hydrolysate formulas, and goat's milk. It is worth considering each of these in turn.

Soya 'milk'

Soya 'milk' is based on protein from soya beans and contains no cow's milk protein, so it is often suggested for a milk-free diet. However, soya milk has a different smell from cow's milk, and children may take some time to accept it. This can be overcome to a great extent if the child drinks from a nursing bottle or trainer cup rather than from an open cup. Soya protein itself can also cause allergic reactions in some children, in which case another alternative should be tried.

Two types of soya milk are available. The first type, *soya formula* milk, is made for babies and younger children. Like *cow's milk formula*, soya formula milk is *modified*, a treatment process by which the balance of protein, fat, carbohydrate, and salt is altered to resemble more closely that found in human milk. It is also nutritionally complete, with a wide range of added vitamins and minerals, including calcium. Several brands are widely available from chemists' shops, including *ProSobee® (Mead Johnson), *Nutrilon Soya® (Cow & Gate), and *Wysoy® (Wyeth). All can be obtained when necessary on a GP's prescription.

The second type, appropriate only for older children, is ready-to-drink liquid soya milk. Several brands are available in health-food shops and supermarkets. Please note that only a few ready-to-drink soya milks contain adequate vitamins or calcium (although they are nutritious in other ways), and are therefore not by themselves appropriate substitutes for cow's milk on a child's diet. However, with an alternative calcium supplement, such soya milks may be acceptable for milk-free diets.

Hydrolysate formulas

Hydrolysate formulas are made by 'predigesting' one of the main proteins in cow's milk protein, either the main curd protein (casein) or the main whey protein (lactoglobulin) during the manufacturing process. This process involves breaking down the whole proteins to peptides (p. 44), destroying the antigenic elements, thus making the formula extremely unlikely to cause allergic reactions. All other components of ordinary cow's milk formula (fats, carbohydrates, vitamins, and minerals) are then re-added and processed to make a modified feed closely resembling human milk in its nutritional value.

Three such formulas are currently available through chemists' shops. They are *Pregestimil® and *Nutramigen® (Mead Johnson), both casein hydrolysate formulas, and *Pepti-Junior® (Cow & Gate), a whey hydrolysate formula. All three are suitable milk substitutes for infants and children, but Pregestimil® is the least appropriate, being designed more specifically for children with intestinal problems. Like soya formula milks, these are unappealing to the adult palate, but most infants readily accept them, perhaps after some initial reluctance. Unlike soya milks, these products very rarely cause allergic responses. They are more expensive than soya formula, but are prescribable by GPs.

Goat's milk

Goat's milk has often been used as a cow's milk replacement. However, soya formula milk or hydrolysate formula are generally preferable, for the following reasons:

1. Goat's milk is not subject to the strict hygiene controls enforced in the production of milk from dairy cattle in order to guard against harmful bacteria. Goats are normally hand-milked instead of machine-milked, adding to the chances of contamination. Furthermore, goat's milk is rarely pasteurized (i.e. heat-treated), while cow's milk is virtually always pasteurized. As a result, when the bacterial content of goat's milk has been checked, it has often been found to contain potentially harmful bacteria. If you do choose goat's milk, and if the packaging does not state that it has been pasteurized, we recommend that you boil it for two minutes. The milk can then be cooled and drunk cold. Unfortunately, boiling does make the flavour worse and reduces the content of some vitamins, especially folic acid, which is already low in goat's milk.

2. Goat's milk is entirely unsuitable for infants under six months of age. Unboiled, it is positively dangerous, for the bacteria it contains may cause a lethal infection in infants. Even when boiled, however, it is hazardous for this age group because it is not 'modified', and therefore lacks the appropriate balance of nutrients.

3. Goat's milk may be difficult to find in retail stores, although availability has improved greatly in recent years. It may be particularly scarce during the kidding season, when it is fed to baby goats. Goat's milk is regularly stocked by health food shops and some supermarket chains, and can be obtained directly from producers in some rural areas. It is available refrigerated or frozen in cartons or in dried form. If you choose goat's milk, we recommend refrigerated goat's milk because it tends to be fresher than frozen milk, and you can freeze it at home yourself for later use. Best of all is dried pasteurized goat's milk.

4. Many of the proteins contained in goat's milk closely resemble those in cow's milk. Therefore, a change to goat's milk may make little difference in a child's symptoms. This phenomenon can be illustrated by an experiment that has been done in guinea-pigs. Guinea-pigs can be made so allergic to cow's milk that 100 out of 100 allergic guinea-pigs will have serious reactions if later given cow's milk. When such guinea-pigs are instead given raw goat's milk, 70 out of the 100 will have reactions. This is a measure of how similar the two milks actually are. If, on the other hand, the goat's milk is first boiled, the number of guinea-pigs having reactions falls from 70 to 30. This is because boiling considerably changes the properties of the proteins in the milk. This is a second reason why it is a good idea to boil goat's milk.

Calcium supplements

Very commonly, children over the age of a year or two will not take any of these milk substitutes, because of their unfamiliar taste. If your child cannot take sufficient amounts of any of them (at least half a pint or 250 ml daily), it is essential to provide adequate amounts of calcium in tablet form, even if the diet contains adequate alternative sources of protein. Perhaps the most suitable are *Sandocal-400® tablets, which are white effervescent tablets. One should be taken daily, dissolved in water or orange juice, and will provide more than enough calcium for most children. There are alternatives; these include *Calcium-Sandoz syrup®, for which the correct daily dose is three teaspoonfuls.

The elimination of cow's milk also means strict avoidance of milk powders (full-fat and skimmed), cheese, cream, butter, yoghurt, non-fat milk solids, caseinates, lactalbumin, lactose, whey, margarines, and shortening containing whey, and whey syrup sweeteners. Food labels should be checked routinely as milk and its derivatives are used fairly frequently in processed foods. Even the labels of tried and trusted milk-free products should be regularly checked as manufacturers may change ingredients from time to time.

Suitable alternatives to butter and whey-based margarines are: Tomor® (Van den Berghs), available from some supermarkets, notably Sainsbury's, and Granose® (Granose Foods Ltd), available from health-food shops. A cow's milk-free soya ice-cream (Soya Health Foods) is also available from many health-food shops.

Implications of eliminating eggs

Eggs are a cheap and nutritious protein-rich food which (unlike milk) can be eliminated from the diet without hazard. They can be easily replaced by foods such as meat, fish, or pulses. However, eggs are an important ingredient of many baked products. Fortunately, there are 'egg replacers' available which can be used in place of eggs (e.g. Egg Replacer®, Nutricia Dietary Products Ltd), but they do not have the same nutritional value as egg. Foodwatch can also supply a 'whole-egg replacer' and an 'egg-white replacer'.

Beware of 'egg substitutes' which contain the egg protein ovalbumin, and are therefore unsuitable.

The elimination of eggs means strict avoidance of eggs, dried egg powder, egg yolk, egg white, egg albumin, egg lecithin, and ovalbumin.

In my hospital, we often ask children who are avoiding egg and milk to avoid chicken at the same time, because some of the proteins in chicken and egg are identical, as one might expect.

Implications of eliminating wheat

Occasionally, avoidance of wheat is advised, though it is in practice one of the most difficult foods to eliminate. It is used in many everyday products—bread, biscuits, cakes, and pasta—and even as a carrier for vitamins and minerals that are added to non-wheat breakfast cereals.

Wheat contributes valuable amounts of protein, B vitamins, fibre, and energy to the diet. If wheat is to be avoided then careful menu planning and advance cooking is essential. Wheat can be replaced by other whole grain cereals and flours, e.g. rice, rice cakes, rice flour, Ryvita-type crispbreads, rye flour, cornflour, barley flour, and potato or buckwheat flour.

Bread, biscuits, and cakes can be made using non-wheat flours, but success normally follows only after numerous attempts at using these flours. The taste will also be very different, although the extent to which it differs depends largely on how much flour is needed in the recipe. Foodwatch produces non-wheat products including a non-wheat bread mix (Versaloaf®) as well as many other dietary products.

The elimination of wheat means strict avoidance of bread (all types containing wheat), bread crumbs (including fish fingers and other foods coated with bread crumbs), hydrolysed vegetable protein (unless known to be non-wheat), rusks (frequently added to sausages), starch (unless known to be non-wheat), wheat bran, wheat binders, wheat flour, wheat germ, wheat gluten, and wheat breakfast cereals.

Other gluten-containing cereals (for example oats, rye, and barley) are suitable unless your child is also allergic to these. Food labels should be checked regularly and, if you are in doubt about any ingredient listed, it is safer to avoid the food.

Implications of eliminating food additives

Over the last few years, most food manufacturers have responded to consumers' demands by minimizing their use of artificial colourings and preservatives. However, there are still many processed foods containing artificial additives, so all food labels must be checked if you wish to ensure that your child's diet will be free of these substances.

Ideally, diets for children with eczema should be based on fresh foods. If convenience foods are used, they should be as free as possible from all artificial colourings and preservatives. These will be listed on food labels either by name or by their 'E' number. Not all E numbers are bad; for example, the colour E160a is carotene, a naturally occurring colour found in plants such as carrots. There are several books and leaflets available which give information on food labelling and E numbers.

The azo colourings and the benzoate preservatives are the two groups of additives most widely considered able to aggravate eczema, but other artificial additives can also do so. There are 11 azo dyes, which I have listed below. The most commonly used are tartrazine (E102), sunset yellow (E110), armaranth (E123), and ponceau (E124). A number without an E before it indicates that the colour is not permitted at all in some EC countries.

- **Yellow**
  E102 Tartrazine
  107 Yellow 2G
  E110 Sunset yellow
- **Red**
  E122 Carmoisine
  E123 Amaranth
  E124 Ponceau 4R
  128 Red 2G
  E180 Pigment rubine
- **Black**
  E151 Black PN
- **Brown**
  154 Brown FK
  155 Chocolate Brown HT

There are 10 benzoate preservatives with E numbers used in food: E210 to E219. Small amounts of benzoic acid occur naturally, but it is the synthetically prepared benzoates that are added to food that more commonly provoke eczema.

Before going on, it might be a good idea to summarize what I have said so far about dietary treatment in atopic eczema. I hope it does not seem too confusing. Foods may be one of the provoking factors that underlie your child's eczema. You may have noticed that eating a certain food makes the eczema worse, but you are in fact more likely to be aware of immediate hypersensitivity reactions to food (p. 39) than of provocation of the eczema itself, which tends to be insidious and unapparent. This means that dietary treatment is worth considering even if there seems to be no obvious link between foods and eczema in a particular child. Because there are no reliable tests to indicate which foods may worsen a child's eczema, the logical approach is simply to try elimination of those foods which have been found overall to be most frequently relevant. Such a diet would be called an empirical elimination diet. The foods that would be most appropriate to eliminate in an empirical elimination diet are eggs and milk, and very often chicken, certain artificial additives, and possibly fish, citrus fruits, and/or wheat. In general, it seems wise not to attempt an elimination diet unless the severity of your child's eczema warrants it, and you should bear in mind that the older the child, the less likely it will be that such a diet will prove beneficial. It is unwise to try this sort of diet without the advice of a trained dietitian, first because you may not succeed without help in completely eliminating these items and, secondly, because elimination diets can be hazardous if their effects on your child's nutrition are not carefully considered.

If an empirical elimination diet is going to help, it will usually do so within six weeks. If it proves helpful, the diet should probably be maintained unchanged for about one year, and, ideally, a dietitian should check on your child's nutrition at least once during this time. At the end of the year, you will need to find out whether your child can now eat these foods without problems.

If a simple diet such as this fails to provide any benefit in spite of being done perfectly, it could mean either of two things. It may be that foods simply aren't relevant in your particular child, or it may be that the relevant foods have not been identified. Generally, if you have tried one or two empirical elimination diets without great benefit, further guesswork is unlikely to be fruitful because of the vast number of foods which may still be playing a role. Unfortunately, the NHS at present provides virtually no specialized allergy service to take the dietary approach further than an empirical diet. Advice on more sophisticated dietary treatments is available in one or two centres, though diets which eliminate a larger variety of foods are in practice only rarely more effective than simpler empirical diets, except perhaps in babies. In babies with eczema who are fed on cow's milk formula alone, dietary treatment will merely involve substitution of the milk formula by either a soya formula or a casein hydrolysate formula, preferably the latter. If a simple empirical elimination diet does not help a baby who is already receiving solids, it may be worth considering a three-week trial of a hydrolysate formula alone. Once again, it is essential that this is not attempted without the help of a dietitian, who will calculate the amount of feed that the baby must take each day. If this proves beneficial at the end of three weeks, other foods can be reintroduced cautiously into the diet in the way outlined previously.

For children over a year old, more sophisticated dietary treatment is very difficult, and the few centres that provide guidance in this area tend to use slightly different methods. Because of the potential nutritional hazards of these types of diet, you should not even *think* about attempting them unsupervised. Occasionally, *elemental* diets have been used. These are complete liquid diets which contain all necessary nutrients in a predigested form to which one cannot be allergic. The trouble is that they taste terrible, even when heavily flavoured. It may be possible to persuade an adult to live on this type of food for a few days, but children quite understandably will not usually have anything to do with it, unless it is given through a tube. Because of these difficulties, elemental diets are generally not used to treat children with eczema, and it is more usual to use a compromise approach, called a *few food* diet. Initially a very limited number of foods is allowed, perhaps only six items. This initial diet is given for two to four weeks. If beneficial, the patient then embarks on a lengthy process of reintroduction of single foods. Each introduction takes a week, as outlined earlier, the idea being gradually to build up a picture of what the individual patient can and cannot eat without aggravation of the eczema. Diets of this type require a great deal of expert supervision, though, despite this, the results of their use have tended to be somewhat disappointing, and they are therefore now only rather rarely recommended.

Conclusions

I am keen to conclude this section on dietary treatment by appealing to readers to keep the whole issue in perspective. Dietary treatment is difficult, not only for parents but also for the child, and great care must be taken to ensure that it does not lead to more suffering than the disease it is being used to treat. Some children who have been treated with diets for many years thereafter become 'faddy' about foods for life, even when they have outgrown their food allergies. It needs to be remembered that dietary treatment has side-effects, just like other treatments, and that these can be serious. In my view, dietary treatment should never be attempted until you are certain that adequate relief of the eczema cannot be obtained by using simple, safe, skin treatments. Only when these have been tried properly and have been proved inadequate should you contemplate an elimination diet. Too many parents are encouraged by well-meaning friends and relatives to cut out milk and other foods almost as a matter of course, even if their child only has the very mildest eczema. This is not justified. Most children with eczema simply do not need such treatment, and many will not benefit from diets. Food allergy is certainly not the only cause of eczema, and many other important elements in the successful treatment of eczema need to be considered first. Having urged prudence, I remain convinced that dietary treatment can be helpful if employed with appropriate judgement and caution.

NON-FOOD ALLERGY

Although foods appear to be of particular significance in generating atopic eczema, we now believe that important roles are also frequently played by allergens other than those in foods. In some cases, it seems likely that non-food allergens may even be the main problem. Most of the important allergens of this type have been identified as a result of research on the causes of asthma (p. 210) and allergic rhinitis (p. 212), but it has gradually appeared more and more probable that they also have a part to play in causing eczema. Asthma and allergic rhinitis are provoked by allergens being breathed into the lungs or nose. This means that in order to provoke asthma or allergic rhinitis, the relevant allergens must occur in a form in which they will be inhaled. In practice, the principal non-food allergens all take the form of particles that are light enough to become airborne. One gets an idea of the size of such particles and of their number in the air when one sees a shaft of sunlight entering a room through a window. These particles land on surfaces in our respiratory apparatus just as they do on surfaces in our homes, where we call them 'dust'. Dust isn't a specific substance, but consists of particles of many different natures. These include pollens, mould spores, tiny fragments of human skin and of skin from any other household animals, and the droppings of the millions of mites that also share our homes.

When these particles are breathed in, very few actually reach the lungs. Most of them stick to the moist walls of the nose and large air passages. The tubular passages which take air down to the lungs are provided with millions of microscopic hair-like projections which move together in such a way that the fluid on their surface is steadily wafted upwards towards the throat, and from there down into the stomach. The same cleansing mechanism is present on the internal lining of the nose. The result is that most of the airborne particles that are inhaled eventually pass down into the gut, just like food. These airborne allergens differ from foods only in the way that they enter the body, and in the total quantities swallowed, which are small. Some antigen will probably enter the blood from particles reaching the intestines, and some, maybe even more, from antigen dissolved on the walls of the nose and the air passages. It is possible that absorption of antigens by these routes is able to provoke eczema, though we don't have definite evidence for this.

In Chapter 5, I described how foods can cause allergic reactions when they come into direct contact with the skin (p. 39). Non-food allergens can almost certainly have a similar effect. Often these are airborne particles which land on the skin just as they land on furniture. However, they probably reach the skin at least as frequently when there is direct contact with surfaces where they are present in large quantities. A good example would be when a child plays on the floor, where there will generally be large quantities of house dust mite droppings and mould spores. Where the skin is already affected by eczema, the moistness of the eczematous skin will serve both to entrap these particles and then to dissolve out the antigens which they contain. The skin surface barrier is impaired in eczematous areas, and here these dissolved antigens will be able to pass through fairly freely. If a contact urticarial type of reaction follows, itching will be amplified, causing more scratching and therefore more damage. Increased surface damage will have the twin effects of attracting more particles and of further increasing antigen passage through the skin. As discussed in Chapter 5, antigens which enter the skin and cause contact urticarial type reactions may go on to cause eczema (p. 42). In this way, non-food allergens may play an important role as contributory causes of atopic eczema. Since contact urticaria and atopic contact eczema appear to involve IgE antibodies (p. 42), skin tests (or RASTs) (p. 43) using these non-food allergens can give clues that they are playing a role in a particular child, and may also help to identify the most relevant ones.

To summarize, many non-food allergens appear able to provoke eczema. Two different mechanisms may be involved, and it remains unclear how important either of these are in practice. Those allergens which become airborne might theoretically exacerbate atopic eczema in the same way as foods can, because the antigens they contain may enter the blood after they have been inhaled; the nose might be an important site for their passage from the air to the blood. These airborne antigens may alternatively contribute to eczema by direct contact with eczematous skin, by provoking reactions of the contact urticarial and atopic contact eczema types. In children in whom these contact reactions are occurring, skin testing and RASTs can be helpful.

It would be valuable at this point to consider individually those non-food allergens which are likely to be most troublesome, and ways that you could try to reduce their potential to worsen your child's eczema.

Pollens

It can be difficult to establish whether pollen is an important provoker of eczema in the individual child. Skin testing can give helpful clues, but unless the test reaction is a very big one, a positive test certainly does not prove the case. The most suggestive feature of pollen allergy is its seasonal nature. As we considered in Chapter 5, in the UK grasses usually pollinate between the beginning of May and the end of July, and trees somewhat earlier—generally between the end of March and the end of May (p. 49). Therefore, regular worsening of your child's eczema during these periods should make you reflect on a possible role for grass and/or tree pollen. In the UK, grass pollen allergy seems to be a greater problem than tree pollen allergy.

The problem with identifying grass or tree pollen as a potential aggravator of your child's eczema is that these substances are exceedingly difficult to avoid. In the case of grass pollen, some simple precautions can help to reduce exposure, but there is no way that you will succeed in eliminating exposure altogether. During the season, make sure that you have no uncut grass in your own garden, and do not let your child play in or walk through uncut grass. When your child plays outside, try to reduce the amount of exposed skin by choosing tops with high necks and long sleeves, and long trousers. Keep your child indoors whenever grass or hay is being cut in the immediate neighbourhood. Windows should, as far as possible, be kept closed, both in the house and when travelling by car.

The ultimate precaution would be to install an electrostatic precipitator to remove pollen particles from the air entering the house. The entire air supply must enter the house through the precipitator, and windows must all be kept sealed to achieve the best results. Such machinery is quite popular in some countries, but is expensive and very rarely used in the UK. It is now possible to purchase air filtration units that are either stationary or portable and which filter the air in the room in which they are situated. Details of these units can be obtained from Bel Air Associates, 63 Oakway, Woking, Surrey GU21 1TS (tel. 04867 3931).

House dust mites

Unfortunately, it is even more difficult to establish whether the house dust mite is playing a role in provoking a child's eczema than it is in the case of grass pollen. Since house dust mites are present throughout the year, aggravation of a child's eczema by them is not likely to be strongly seasonal. As in the case of grass pollen, a skin test or RAST may be helpful. A strongly positive test indicates that direct skin contact with the mite and its droppings would be able to cause contact urticaria or atopic contact eczema, if suitable exposure occurred.

An attempt to reduce exposure to house dust mite allergens is worthwhile in any child whose eczema is a great enough problem to warrant it, and where there is any reason to suspect a causative role for them. In fact, because it can be so difficult to establish such a role, it makes sense to undertake at least minimal anti-mite measures in any household where there is a child with eczema. The more difficult the eczema is to control by other means, the more effort should be made to eradicate the house dust mite and to reduce your child's exposure to dust. However, you do need to find a reasonable compromise in terms of just how far you take your efforts.

The following measures represent a sensible regime that is not too demanding. Anti-mite efforts should be concentrated on your child's bedroom, and on the room in which your child spends most time when at home during the day, probably your living room.

Mattress and pillows should be enclosed in microporous covers which do not allow through house dust mite droppings, but do allow air and moisture exchange. Such covers are far more comfortable for your child than the plastic sheeting that used to be used for this purpose in the past.

Duvets are a good idea, but choose a man-made filling. Similarly, choose man-made fillings for pillows. Wash the duvet and pillow frequently—say weekly—and the sheets and pillowcases every one or two days. If your child prefers blankets, choose acrylic or cotton ones because these are easier to wash and dry than woollen ones.

Details of suitable covers and other bedding products can be obtained from:

1. W. L. Gore and Associates Ltd, Church Gate, Church Street West, Woking, Surrey GU21 1DJ (tel. freephone 0800 515730).
2. Medivac Healthcare Products, Bollin House, Riverside Works, Manchester Road, Wilmslow, Cheshire SK9 1BJ (tel. 0625 539401).
3. Slumberland p.l.c. Medicare Division, Salmon Fields, Oldham OL2 6SB (tel. 061 628 5293).
4. Allerayde, 147 Victoria Centre, Nottingham NG1 3QF (tel. 0602 240983).
5. Alprotec, Advanced Allergy Technologies Ltd., Royal House, 224 Hale Road, Altrincham, Cheshire WA15 8EB (tel. 061 903 9293).

If possible, give your eczematous child a separate bedroom, as this will reduce the house dust mite population and the amount of dust raised into the air. In any case, do not let an eczematous child sleep in the lower bunk of bunk-beds.

Choose washable linoleum or vinyl flooring whenever possible rather than carpets, and do not use scatter rugs. In the bedroom, use plastic material for curtains rather than fabric. Chairs should be of painted wood or plastic, and not upholstered. It is unlikely that you will be keen to apply these principles in the rest of the house unless your child has severe eczema, or asthma as well. Bear them in mind, none the less. Try to minimize the number of cuddly toys, unless these are merely on view in a closed display case. Cuddly toys are fine, though, if you only allow washable ones, and wash them every week.

Clean rooms as often as possible, but remember that dust mite droppings are likely to be more of a problem if they are made airborne by being disturbed. Therefore, wipe down surfaces with a damp cloth rather than a duster. Since house dust mite droppings may simply pass straight through the bags in conventional vacuum cleaners, vacuum cleaning may make the situation worse. It is important therefore to use one of the machines that is designed to trap the great majority of these droppings; good examples are Medivac (Medivac Healthcare Products, Bollin House, Riverside Works, Manchester Road, Wilmslow, Cheshire SK9 1BJ, tel. 0625 539401), Nilfisk (Allerayde, 147 Victoria Centre, Nottingham NG1 3QF, tel. 0602 240983) and Vorwerk (Vorwerk UK Ltd, Toutley Road, Wokingham, Berks RG11 5QN, tel. 0734 794753). A cheaper alternative, but not as effective, would be to fit a special filter to your existing vacuum cleaner; such filters are available from Allerayde (address above). Vacuum clean carefully and as frequently as you can manage. Remember to vacuum the surfaces of settees and mattresses (having removed the microporous cover if one is in use), paying special attention to seams. After cleaning, any particles that have been lifted will take a little time to settle. For this reason, you should keep your child well away when you are cleaning and, if dust has been disturbed, keep your child out of that particular room for a few hours.

I have always been suspicious that eczema in certain areas may be particularly suggestive of a special role for the house dust mite. Those sites are the front of the knees, perhaps due to kneeling on carpets, the backs of the thighs, perhaps due to sitting on settees, and the cheeks in older children, in whom facial eczema is relatively unusual, perhaps reflecting contact with mite antigen on the pillow. In these areas, the problem would be the combination of contact with house dust mite droppings and friction. If your child does sit on upholstered chairs or play on carpeted floors, have an old sheet that can be placed under them and be regularly washed. We have already considered ways of treating the pillow.

Although dry air is disliked by the house dust mite, I do not agree with the advice sometimes given, that dehumidifiers should be used. The problem is that excessively dry air will almost certainly aggravate your child's eczema, while the mites will continue to proliferate happily in more humid micro-environments. Nevertheless, rising damp and other sources of unwanted moisture should be dealt with. In respect of house dust, ducted air heating tends to make matters worse by constantly disturbing the air and raising the dust particles.

Interest has recently been renewed in the possibility of eradicating house dust mites by using chemicals, technically called *acaricides*, because of the availability of treatments which are relatively non-toxic to humans. There are however many problems, particularly problems of getting the chemicals to the mites in sufficient concentration. It is also important to remember that these chemicals may kill mites, but will not help your child unless all the faecal particles already present are removed before the mite population recovers. This means that their use must be accompanied by vigorous cleaning along the lines already described. One of the most promising of those currently available is +Acarosan®, which can be obtained through Crawford Chemicals, Denbigh House, Denbigh Road, Bletchley, Milton Keynes MK1 1YP. This consists of benzoyl peroxide, which kills mites but appears to be safe for children and pets, and a special polymer to which the house dust mites droppings stick so that they are trapped in large aggregates and can then be more effectively gathered by vacuuming. Acarosan® is applied to carpets in the form of a powder, and to mattresses, soft furnishings, and cuddly toys in the form of a foam. A single application of Acarosan® appears to be effective for about three months, much longer than most other acaricides. Long-term safety has not yet been firmly established for any of the available acaricides, and neither has it yet been proven that their use can help children with eczema. Nevertheless, this is an important area of research.

Another chemical that appears able to reduce exposure to faecal antigen is tannic acid (+Banamite® Spray, available from Medivac Healthcare Products, Bollin House, Riverside Works, Manchester Road, Wilmslow, Cheshire SK9 1BJ, tel. 0625 539401). This chemical can denature mite faecal antigens and pet dander antigens, and is sprayed on to carpets and furnishings. It has been shown to reduce dust mite antigen levels, but it remains to be demonstrated that its use can lead to an improvement in atopic eczema.

Although desensitizing injections may occasionally be helpful for eczema caused by grass pollen, the same has not so far been shown to be true for house dust mite allergy (see p. 51). Nevertheless, it is certainly possible that such treatment would help some children, and this is an area in which more research is needed.

Pets and other animals

Bluntly, it is very unwise to have a warm-blooded pet such as a cat or dog in your home if your child has eczema. In practice, many families have pets which became family members before the child with eczema, and parents are loathe to part with them for this reason. Also, children with eczema are often very keen on pets, just like any other children, and parents do not want to remove such a child's pet when they are aware that the child is missing out on other pleasures. But perhaps the most important reason why parents hold on to pets is that they may lack any clear evidence that the pet makes their child's eczema worse. Interestingly, many parents only notice that their child is allergic after the pet has been removed, when occasional contact with the same or another pet may cause obvious reactions. Some parents observe allergic reactions when their child has contact with other people's pets but apparently not their own. This phenomenon is very real, and may be due to some kind of desensitization to the family's own pet. However, it is my suspicion that though this desensitization may suppress the more obvious urticarial type of reactions, it may not stop the pet aggravating the child's eczema in a more insidious and covert way.

Another factor that needs to be considered is the likelihood that contact with pets increases a child's risk of developing asthma.

Though I am myself fond of cats and dogs, I would advise families not to have these pets where there is a child with eczema. Certainly you should not take one on, and if you have one already, you should seriously consider removing it from your home. The best way is to look out for a friend or relative who is prepared to look after it, to save you having to put it down. There would then be no problem with your child seeing the pet from time to time. You should be aware that it may take several months for you to see any improvement in your child's eczema after the pet has left. You will also need to vacuum and clean very thoroughly.

If you decide to keep your pet, make sure you *never* allow it to sleep in the same room as your eczematous child, and do everything you can to ensure that it doesn't ever go into the child's room.

Contact with horses should also be minimized in children with eczema. If children do ride despite this advice, they should not be involved in cleaning out or grooming activities. If anyone else in the family rides, they will need to be careful with the clothes and footwear in which they do so. It is best for them to change before coming into the home, and keeping these items well away from the child with eczema.

Allergies to virtually all mammals have been recorded, both to pet species such as rabbits, mice, and guinea-pigs, and to farm animals such as sheep, pigs, and cattle. Children with eczema should have as little as possible to do with all such animals, and you should be aware that many antigens are common to different mammals. Animals kept outside, like rabbits, are not such a great problem, but your child should be discouraged from handling them or cleaning them out (this shouldn't be too difficult!).

Allergy to birds is likely to be a similar problem. It is the dander and droppings that seem to be responsible rather than the feathers themselves.

If your child is desperate to have a pet, go for something like tropical fish in preference.

Moulds

Though very little is known about the role of mould allergy in eczema, the possibility that it may contribute to your child's problems should be considered. Some precautions make sense. Eliminating excessive damp from the house helps to reduce both the amount of mould and the number of house dust mites. You could consider having your home inspected by a timber preservation specialist, to make sure you do not have dry rot and to measure the amount of rising damp in your walls.

You should avoid storing large quantities of foods such as grain, beans, dried fruit, and apples.

Out of doors, avoid assembling large heaps of compost, and keep those you do have covered with polythene. Collect up fallen fruit and leaves, and dispose of them. Discourage your child from playing in barns or other similar outbuildings where moulds will flourish.

ANTI-ALLERGY DRUGS

In Chapter 3, we considered the reaction that occurs when an antigen encounters its specific IgE antibodies on the surface of a mast cell. The cell is triggered to discharge substances that cause inflammation, including histamine (p. 10). Almost all children with eczema are atopic, i.e. they produce excessive amounts of IgE antibodies against common antigens. Many children who do not have eczema are also atopic; some will have asthma, some will have hayfever, but some will be entirely well. All these atopic children have mast cells that are highly sensitive and liable to discharge their contents readily on exposure to the appropriate antigen. As we have discussed, this sensitive (i.e. allergic) state is detectable by skin testing (p. 43) and by the IgE RAST (p. 43). Mast cell sensitivity in the lungs plays a part in causing asthma, and mast cell sensitivity in the nose and eyes plays a part in allergic rhinitis and allergic conjunctivitis. In the skin, mast cell sensitivity is important in contact urticarial reactions (p. 39), and appears to play a part in the reaction called atopic contact eczema (p. 42). In general though, its role in atopic eczema is unclear.

Drugs have been developed which are able to reduce discharge of mast cell contents after antigen has met antibody on the mast cell surface. For this reason, these drugs are often referred to as *mast cell stabilizers*. The best known, and to date the most effective of these, is known as sodium cromoglycate. This drug has proved helpful in many cases of asthma, for which it is given by inhalation in a form called 'Intal®. Other preparations of sodium cromoglycate called *Rynacrom® and 'Opticrom® are often useful when given respectively in the nose and eye for allergic rhinitis and allergic conjunctivitis.

Sodium cromoglycate has also been tried in eczema by applying it to the skin in the form of an ointment, but does not appear to be effective when used in this way. It is not clear whether this is just because insufficient amounts of the drug succeed in getting through to the mast cells because of the skin's barrier effect. Alternatively, it may be that sufficient does get through, but then fails to help simply because mast cells are not responsible for eczema. However, there is evidence that mast cells in the skin are not stabilized by sodium cromoglycate to nearly the same degree as those in the lungs, nose, and eye.

If sodium cromoglycate is given by mouth, very little enters the blood, almost certainly insufficient to have any effect on distant organs like the skin. Given this way, however, it may be able to exert a local effect on the lining of the intestines. I mentioned in Chapter 5 (p. 38) that children with eczema have intestines which appear to allow the passage into the blood of increased amounts of food antigens, and that this might reflect damage to the wall of the intestines by an allergic reaction to foods. This allergic reaction would be likely to involve IgE antibody and mast cells. It would follow that if this initial reaction in the intestines could be prevented, one might be able to prevent the eczema which appears to be its consequence. Later I will discuss the possible role of this mechanism in the effectiveness of Becotide® given by mouth (p. 197). Some early studies suggested that sodium cromoglycate, in the form designed to be given by mouth ('Nalcrom®) might be helpful in atopic eczema. Unfortunately, it now seems that this early optimism was largely unjustified, and that in general Nalcrom seems to have little beneficial effect in children with eczema. Why not is unclear, but this is a familiar problem in medicine: a drug does not work when theoretically it should. More research is needed to discover why this is. It seems possible that Nalcrom will help a small proportion of children, perhaps those with more clear-cut aggravation of their eczema by foods, but this has not been clearly proven.

New drugs of this type are being developed which will need careful appraisal in atopic eczema. Most of them differ from sodium cromoglycate in being absorbed in reasonable amounts into the bloodstream. Whether this will make them more effective or only more toxic remains to be seen. Even so, there is room for guarded optimism that some of these drugs will prove helpful.

DESENSITIZATION THERAPIES

It would seem logical and desirable, since adequate allergen avoidance is so difficult to achieve, to seek ways of reducing the sensitivity of the allergic person to particular allergens instead. Treatments having this aim could be considered collectively under the heading of *desensitization* techniques.

There are three techniques that need to be considered: *desensitization by allergen injection, sublingual desensitization*, and *enzyme-potentiated desensitization*. In each case, the strategy is to administer a really minute amount of the relevant allergen, so minute that an allergic reaction is not triggered. Thereafter, the dose administered is very gradually increased, and it turns out from experience that the patient will tolerate this process until a substantial dose can be given without any ill effect. When the dose tolerated is similar to the level of exposure that is likely to occur naturally then desensitization has been achieved. Experience has also shown that this state of tolerance or desensitization can be maintained indefinitely either by occasional booster administrations of allergen or by natural exposure.

Conventional desensitization by allergen injection

A good example of the appropriate use of desensitization is the desensitization treatment used to treat bee or wasp sting allergy. This is a highly dangerous form of allergy that may put individuals at risk of death if they are stung. This type of desensitization is administered by jections into the skin. Initially very small doses of bee or wasp venom are injected, and successively larger doses are given until the dose is close to the amount that would be injected by the insect itself. This works very well, and undoubtedly saves lives. It does, however, carry a risk from allergic reactions to the injections themselves, sometimes serious, and this type of treatment therefore has to be given by experts in a centre which is suitably equipped to deal with potentially serious allergic reactions should these occur.

Desensitization injections are now hardly ever used in the UK except for bee and wasp venom allergy, because the view has been taken that overall the risks outweigh the benefits. The treatment is, however, still in widespread medical use in other countries as a treatment for a wide variety of allergic disorders, particularly nasal allergy to pollens. It has also been used to treat house dust mite allergy in patients with asthma. Debate continues as to the benefits and the risks of this type of treatment.

In the past, when we used injection desensitization more freely, I gained the impression that children who had experienced provocation of their eczema during the grass pollen season could be helped by desensitizing injections. This treatment used to be given immediately prior to the grass pollen season. We have also undertaken trials of house dust mite desensitization for children with atopic eczema. Although these trials did not establish that this treatment was of value, we formed the opinion that prolonged courses of injections might be of value. Unfortunately, we were unable to continue with this research because the tide of medical and legal opinion turned against desensitization treatment by injection.

Sublingual desensitization

Allergens are absorbed into the blood through the lining of the mouth underneath the tongue (the *sublingual* route), and the effect on the immune system is somewhat different from the effect of injected allergen. There is some evidence that desensitization by this route can be achieved safely and with relatively small numbers of doses, even perhaps single doses, but research is needed to establish that this form of desensitization would be effective in atopic eczema.

Enzyme-potentiated desensitization

This is a special type of desensitization in which very small allergen doses are administered together with an enzyme called *beta-glucuronidase*. The enzyme is believed to allow extremely minute doses of allergen to provide desensitization with greater safety, and there is some evidence that this technique may allow desensitization with very few doses, even perhaps with a single dose.

The treatment has most often been given by the 'cup' method, in which the desensitizing mixture is contained in a chamber placed open side down on the skin. The skin is scratched to allow slow penetration of the mixture and the chamber is left in place for 24 hours. An alternative is to use an intradermal injection (into the *dermis*).

An important potential advantage is that mixtures of many antigens can be used safely, and it may therefore not be necessary to identify a patient's precise allergies prior to therapy. This would be a great help in the case of a disease like atopic eczema, in which the identification of relevant allergens is so difficult. Further investigation of this technique would be very worthwhile.

AVOIDING IRRITANTS

We considered the harmful effects of a wide variety of irritants in Chapter 5 (see p. 57). The mechanism by which such irritants contribute to eczema has nothing directly to do with the immunological system, nor therefore to allergy. Although irritants are undoubtedly important aggravators of atopic eczema, they are not capable of causing it on their own. It seems probable that the disease is always initially provoked by allergy, though the responsible allergens will differ from child to child. Irritants, on the other hand, appear to be very important in maintaining activity of the disease, and in many cases, irritants eventually seem to play the major role.

Careful observation will often give strong clues to which irritants are most likely to be causing problems to your own child. Some are easily avoided, others only with difficulty. The following is a general guide to reducing exposure of your child to the principal irritants.

SOAP AND DETERGENTS

Avoid *soap* and *detergents* when washing your child, similarly soapy additives such as foaming agents, or *bubble baths*. If you or your child ever *have* to use soap, use a non-alkaline soap such as +Neutrogena® or +Dove®. An alternative is a non-soap cleanser such as +Aveenobar®, or *solid creams*. The *solid creams* are likely to appeal particularly to teenagers; suitable versions for children with eczema are made by some of the cosmetic companies such as RoC. They are designed to be used as cleansing agents, and look like a bar of soap, but do not in fact contain soap at all.

Use a non-enzyme detergent for washing clothes. I feel that machine-washing is best, because rinsing tends to be more thorough. There are a number of suitable alternative detergents easily obtained from supermarkets.

PROLONGED IMMERSION

Discourage your child from playing with water at all. Do not let older children do the washing up, or other wet work. If they insist, they should wear vinyl gloves (not rubber, because of the risk of allergy), with cotton gloves underneath; these are widely available in pharmacies and supermarkets, or by mail order from Bio-Diagnostics Ltd, Upton Industrial Estate, Rectory Road, Upton upon Severn, Worcs WR8 0XL (tel. 06846 2262).

Play with sand is also best avoided.

SWIMMING

Swimming, particularly in swimming pools, has a drying effect on the skin. However, if your child is really keen to swim, or you are really keen to take your child swimming, do so, but with adequate precautions. Before swimming, cover the skin with an oily barrier to protect it; white soft paraffin (Vaseline®), or a mixture in equal parts of white soft paraffin and liquid paraffin, are ideal for this purpose. As soon as possible afterwards, give your child a bath or shower with appropriate moisturizers, as described on p. 96.

HARD WATER

Consider using soft water to wash or bath your child. There is some evidence that atopic eczema is commoner in those parts of the country where the water is hard, and many parents have told me that buying a domestic water softener has had a beneficial effect on their child's skin. Exactly why *hardness in water* should be an irritant is unclear. To soften your child's bath water does not necessarily mean that you have to invest in a complete domestic water softening system costing hundreds of pounds. Water softening equipment is available from a number of different manufacturers. Permutit softeners have an established reputation and you should be able to obtain a 10 per cent discount if your child has eczema (contact: Ecowater Systems Ltd, Unit 1, The Independent Business Park, Mill Road, Stokenchurch, Bucks HP14 3TP, tel. freephone 0800 521143).

COSMETICS

Few teenagers with eczema find that they can tolerate much in the way of *cosmetics* other than moisturizers. However, certain companies do undoubtedly produce superior products for irritable skin; these include RoC® and Clinique®

FOODS

Don't let your child handle *irritant foods* if the hands are affected. This means fruits, especially citrus fruits, raw vegetables, especially onions and tomatoes, and foods that are salty or spicy. Peel oranges yourself. If your child is young and a messy eater, avoid giving foods that will irritate the face; these include all the above and there are certain special offenders such as tomato ketchup and tomato sauce, and Marmite.

WOOL AND NYLON

When dressing your child, avoid having *pure wool* or *nylon* next to the skin. Wherever possible, any clothing that comes into direct contact with the skin should be pure cotton or a high cotton mixture. A good range of 100 per cent cotton clothes is obtainable by mail order from Cotton-On, 29 North Clifton Street, Lytham FY8 5HW (tel. 0253

736611). There are numerous other firms that sell cotton clothing; these include Kids' Stuff, 10 Hensmans Hill, Bristol BS8 4PE (tel. 0272 734980). Other firms will make cotton clothes to order, including 100 Per Cent Cotton, 22 Hambledon Court, Holmwood Gardens, Wallington, Surrey SM6 0HN (tel. 081 669 6028).

Remember that wrists and necks are likely to come into contact with pullovers, whatever is worn underneath, and these areas can suffer as a result. For the same reasons, try not to let your child's skin come into direct contact with carpets. It is a good idea to place a cotton sheet over the carpet where your child is playing.

OVERHEATING

Do not overheat your child. Remember that *heat* is one of the greatest enemies of anyone with eczema.

It is worth looking at this problem more closely. Human beings need to maintain a constant body temperature. Heat loss through the skin must therefore be precisely matched by heat production. Heat is produced mainly by the liver, but also as a by-product of muscle activity. This system has both a coarse and a fine control. Coarse control is the function of thyroid hormones. If the body's temperature falls, thyroid hormone production is stepped up. Thyroid hormone works by stimulating the liver to increase its heat output. Nevertheless, a fine control is needed in addition; if one had to depend entirely on the thyroid gland, there would be a considerable lag between any change in body temperature and the resulting change in liver heat output. This fine control must therefore provide for a more rapid response. A sudden fall in body temperature may be compensated for by shivering, an involuntary mechanism that takes advantage of the heat produced by muscle activity. If the body's temperature rises rapidly, heat loss can be increased by allowing more blood to flow close to the surface of the skin, where it will be cooled, and by sweating; the evaporation of sweat produces swift cooling. A potential problem however is that heat loss by both these mechanisms can be compromised both by clothing and by high air temperatures.

This whole system bears a close resemblance to domestic central heating. The boiler is the liver, the circulating water is the blood and the pipework is the blood vessels. Coarse control is achieved by turning the boiler up or down. When it comes to fine control, the equivalent of shivering is turning on an electric fire. The windows are like the blood vessels in the skin; open them and heat loss is accelerated; close them and it is reduced.

In most people, this system is fairly finely balanced. But in children with extensive eczema, it can become disturbed. The problem is caused by the huge heat losses that occur through inflamed skin, particularly when the area affected is large. As a result of this inflammation, large volumes of blood flow very close to the surface, which is why the skin is red. To compensate for such high losses, the liver has to produce more heat. This is much the same as having all the windows in your house open on a winter's day; you need to turn the boiler up to keep the house warm. In turn, this would need extra fuel, which is why children with severe eczema tend to have good appetites. They are making heat very much faster than other children, but as long as they are also losing it faster, their condition remains stable. However, if the heat loss is in any way prevented, the child will quickly overheat. This may happen if the air temperature is high, if they are wearing too many clothes, if a bath is too hot, or if they exert themselves physically. It may even happen if they have a hot meal. This is exactly what would occur if the boiler had been turned up to compensate for the windows being open, and you suddenly shut all the windows.

This unsatisfactory and unstable situation is aggravated by the fact that sweating tends to be impaired in eczematous skin, so that this emergency system for cooling the body is less efficient. The inflammation appears to damage the ducts that lead up from the sweat glands through the epidermis to the skin surface. As well as causing blockage of these ducts, the eczematous process seems to make these ducts leaky, as a result of which some sweat seeps out into the already inflamed tissues. This seepage of sweat causes further irritation. As a result, when the child overheats the skin may become unbearably itchy.

I hope that all this helps to explain some of the problems of heat regulation in children with eczema. Your child needs to be able to lose the extra heat that is being produced. To help prevent overheating, rooms should be kept as cool as is comfortable. Your child's bedroom should in particular be kept cool, preferably no more than 60°F (16°C). Baths should be cooler than you would like yourself. Be very careful not to overdress your child—rather slightly too cold than too hot.

FOOTWEAR

Pay careful attention to *footwear* if your child has eczema on the feet; this subject is dealt with in detail on p. 145.

FINGERNAILS

Cut your child's *fingernails* regularly. Scratching is, in a sense, the most ferocious irritant of all; keeping the nails short helps to reduce the damage. Use a sapphire or diamond file (available from all chemists) rather than scissors, clippers, or an emery board. The best plan is to use the file every day or two for just a few minutes to make sure that there are no sharp edges. Wearing mitts at night provides further protection.

CLIMATIC EXTREMES

Protect your child's skin from extremes of climate. *Extreme cold* is a powerful irritant to skin. Cover up as much of your child's skin as possible before going outside in cold weather, particularly when it is windy. Gloves are especially important. Put a good layer of a moisturizer such as white soft paraffin (Vaseline®), or a mixture in equal parts of white soft paraffin and liquid paraffin, on any exposed skin as this will provide some protection.

As we have already considered, *extreme heat* is also very irritating; therefore dress your child lightly in hot weather, and discourage too much rushing around when it is hot.

SUNLIGHT

Occasionally direct *sunlight* helps eczema, but more often it proves irritating. You should use good sun protection on your child with eczema. This means high necks, long sleeves, and long trousers wherever possible. Dark-coloured fabrics tend to give better protection than light-coloured ones. Suitable non-irritant sun protection creams for children with eczema include *RoC Total Sunblock® cream (SPF 15) and *Sun E45® lotion (SPF 15).

LOW HUMIDITY

Consider the use of humidifiers. The air inside homes in colder climates tends to be dry in the winter, especially when it is very cold. Central heating greatly exacerbates the situation. While cold, dry air has a bad effect on the skin in general, on eczematous skin it can be very detrimental. Obviously you don't want conditions to be frankly damp, but gentle humidification can help a lot in the depth of the winter, especially in your child's bedroom, and if you can afford it, in your living room also. The best way of increasing humidity is to use electric humidifiers; the type of humidifier that is hung on radiators is unlikely to be adequate. If you want to take this seriously, buy a *hygrometer*, which is an instrument that measures humidity; *hair hygrometers* are probably the most suitable type for home use. Try to keep the reading between 50 per cent and 60 per cent. Probably the most economical form of electrical humidifier is the evaporation type; these are ideal, and are best purchased with a *hygrostat*, which will help maintain humidity within the desired range and avoid overdoing it. Evaporation humidifiers are available from a variety of sources, including the Air Improvement Centre, 23 Denbigh Street, London SW1V 2HF (tel. 071 834 2834).

OTHER APPROACHES TO TREATMENT

HOMEOPATHY

Homeopathy is an approach to medicine that was developed in Europe during the last century. Unlike other forms of complementary medicine, it is established as a specialty in the UK within the NHS. There are two cardinal features which distinguish it from mainstream medicine. The first is the principle of treating 'like with like'. The basis of this is the concept that a substance that produces symptoms in a healthy person can cure the same symptoms in a sick person. In practice, this is interpreted fairly freely. Thus, jaundice, in which the patient goes yellow, is often treated with gold. Eczema, an 'eruption' of the skin, is frequently treated with sulphur, a product of volcanoes.

The second feature of homeopathy is that the remedies are given in minute quantities. The aim is to give an infinitely small dose. Such minute doses do, of course, have the great virtue of being harmless, and the reason that homeopathic medicine first caught on in the 19th century has quite a lot to do with this freedom from hazard. At a time when conventional medicine could do little for patients, and when medical remedies very often proved even more unpleasant than the diseases being treated, homeopathy was safer and more attractive. However, though I applaud the gentler approach, I must confess I find myself unconvinced by the principles of homeopathy, and unimpressed by its success as a treatment for troublesome eczema.

HYPNOSIS

The medical use of hypnosis should not be confused with the routines used for entertainment on the stage. A variety of techniques are used medically, all of which have their basis in a belief that the mind can be harnessed to help patients overcome the effects of their disease. Whereas it is generally acknowledged by the medical profession that psychological factors can aggravate illness, it is much less widely acknowledged that the converse might therefore be true, that the mind could be used positively to fight disease. The failure of modern medicine to exploit the power of the patient's mind is, in my view, one of its greatest weaknesses.

Relaxation techniques can be used to treat any disease which is aggravated by mental arousal. A good example is high blood pressure. It has been shown that most people can lower their own blood pressure by a conscious effort to relax mentally and physically, and a similar method can be used to treat duodenal ulcers by reducing the stomach's secretion of acid. In eczema, the symptom of itching is highly susceptible to psychological influences. The therapist aims to teach the eczema sufferer ways of distracting the mind from the skin. Again there are many ways of doing this. In children, a good technique is to use the child's powerful imagination to create a mental image of a peaceful happy environment which will keep the mind absorbed long enough to allow a bout of itchiness to abate. The aim would be to teach the child to use the technique alone, or with the help of a parent; this is called *self-hypnosis*. I recall one girl who liked to imagine that she was in a landscape covered in snow. The cold made her feel less hot and bothered. In the landscape there was a friendly snowman, who would cuddle her. Her parents told me that her skin would become noticeably cooler while she was doing this. While she was deeply involved in this imaginary world, she would appear to be asleep, but would wake up within a few minutes, feeling relaxed and free from the itch. Of course, not all children can do as well as this, but most children over the age of about seven years can be helped if they want, and if parents are sympathetic to the method. This self-relaxation technique can be especially useful as an aid to getting to sleep, and has the distinct advantage over sedative drugs like antihistamines that there is no hangover the next morning. Another trick that can be used is for the child to imagine that their itch can be 'channelled' into one part of the body, say the little finger. Then, by squeezing this finger, the symptom can be controlled. There is no doubt that these approaches can be useful, and it is of course particularly satisfying for eczema sufferers to feel they can get some control of the disease themselves.

Post-hypnotic suggestion is the technique that most people associate with hypnosis. The therapist has first to get the patient into a deeply relaxed trance-like state, when an idea or instruction can be planted into the patient's mind, with the aim that it will be triggered under specific circumstances for a period afterwards. This is the stuff of the stage hypnotist and, although used in medical hypnosis to make smokers feel sick whenever they think of lighting up, for example, it has never been demonstrated to work well in practice. In the case of eczema, therapists will occasionally use a suggestion such as, 'if you feel itchy in school, you must sit quietly, look down at your lap, and then gently put one finger on the bit of your body that is uncomfortable; doing this will help the itch go away'.

ACUPUNCTURE

Acupuncture is an integral part of the traditional Chinese system of medicine, which has developed over many hundreds of years. According to the Chinese, there is an invisible energy force, called '*Qi*', which flows between the limbs and the vital internal organs through channels called *meridians*. The Chinese believe that all illness reflects a disturbance of the natural harmony of the body, and that acupuncture helps to correct this, because the insertion of needles into the channels can modify the flow of *Qi* to the various organs. Acupuncture does not always require the use of needles, and in children alternatives that are often preferred include the application of pressure, heat, or a laser beam.

The value of acupuncture in the treatment of pain is widely recognized, though the way it works is unknown. It has been suggested that it may reduce perception of unpleasant sensations such as pain by altering the levels of certain natural body chemicals known as endorphins. These are chemicals which have effects similar to morphine, the most potent of all known pain-relieving drugs.

Benefit in the treatment of atopic eczema is not established. When I visited China recently I learned that acupuncture is in fact used rather rarely to treat eczema, or any skin disease, and it is probably relatively rarely that it will help.

IONIZERS

Ionizers emit negative electrical charges into the air. These negatively charged ions are claimed to have generally invigorating properties. They are also claimed to have an air-cleaning effect; ionization is said to cause allergenic particles in the air to stick to nearby room surfaces so that they are removed from the air. As far as I am aware, these claims have not been scientifically proven. While the air-cleaning effect might be helpful for a disease such as asthma in which the disease can be provoked by inhaled allergens, in atopic eczema direct contact with non-food allergens appears to be the greater problem, and this would be little altered by ionizers.

HOSPITAL ADMISSION

When eczema gets out of control, the question often arises whether a child should be admitted to hospital. In the short term, hospital admission certainly may be beneficial, both to the affected child and to the family. When weighing up the pros and cons of admitting a child with eczema, most dermatologists and paediatricians will consider the interests of the rest of the family as much as those of the child himself. Living with a severely eczematous child who is getting steadily worse in spite of all efforts can be an extremely harassing experience for parents, as many readers will know. Faced with the unrelieved suffering of their child, the hopelessness felt by parents is frequently heightened by lack of sleep and by disagreements about treatment. Brothers and sisters feel neglected, and may be disturbed by their parents' obvious distress. This distress in its turn becomes a major factor in further aggravating the eczema of the affected child, and the family can become caught up in a vicious spiral. It is in these conditions that admission can be most effective. The child with eczema is taken out of what has become a stressful home situation into the relatively relaxed atmosphere of hospital, and is no longer the focus of such intense anxiety. The family can rest and recover their sapped strength.

Parents are often amazed at how effective hospital admission can be in clearing up a bad attack of eczema. Indeed, it can be upsetting to see the treatments that failed at home now appearing to work so well. The reason for the improvement in the eczema probably owes less to the treatment itself than everybody thinks, including the doctors and nurses. As I have already suggested, the most important influence is often the change of scene itself, and the child's removal from what has—through nobody's fault—become an electrified atmosphere at home. Though you may think this sounds harsh and old-fashioned, experience suggests that the benefits of admission, to the eczematous child and to the family, can be lessened if parents spend too much of the period of admission in hospital with their child. It is an important opportunity for everyone to get a break from one another and rebuild their reserves. It is particularly an opportunity for parents to fortify their relationship with each other, and, very importantly, with their other children. Visiting is obviously important, but each visit should be short, no more than an hour or so. This is a generalization, of course, and there will inevitably be times when a child with eczema is admitted and the parents' presence is more important.

There may be other reasons why a spell in hospital can help a child with eczema. Levels of house dust are generally low, for example, and there are no animals. However, although a short stay in hospital can for all these reasons be a good idea as a temporary measure, it can never become a substitute for good home treatment. One hears of children who spend more time in than out of hospital; this is a bad thing. Often the treatment regime used in hospital is of the 'sledge-hammer' type; for example, very potent topical steroids may be applied. Although occasionally this may lead to lasting improvement, more often it does not, and the eczema merely deteriorates again as soon as the child goes home. This can be quite demoralizing for parents. I prefer to see admission used in a way that is designed to have a more lasting effect, as an opportunity to identify an appropriate treatment routine for use at home.

TREATMENTS FOR EXCEPTIONALLY SEVERE ECZEMA

There is a small range of treatments kept in reserve for children whose eczema has become exceptionally severe, and has persisted in being severe despite conscientious application of the treatments we have already considered. Parents often ask me, 'how severe is our child's eczema?'. This question is an extremely important one, but it cannot be answered accurately if one only takes into account the visual appearance of the eczema. It is much more pertinent to judge the severity of eczema by considering the handicap it causes. School attendance and performance need to be taken into consideration, as does the child's ability to take part in sporting and social activities. One needs to assess the impact of the disease on that more subtle measure we call quality of life. The impact of the child's disease on the family can be very revealing in this respect.

The more the child's quality of life is diminished by the eczema, the more urgent it becomes to intervene with effective treatment. Effective treatment is essential not just because of the immediate suffering caused by the eczema, but because unrelieved severe eczema may result in irreversible long-term harm. Perhaps the clearest example would be the long-term damage that eczema can do to a child's education. It is more difficult to assess the long-term social and emotional damage that eczema do, but there can be no doubt of its ability to cause these types of effect.

The following are the principal measures that need to be considered in children whose eczema is severe and is causing substantial handicap, despite sustained efforts to improve the situation with less potent treatments.

PREDNISOLONE

Prednisolone and its close relative *prednisone* are the synthetic steroid hormones that are most often given by mouth to treat severe eczema. We considered some of the properties of steroids earlier in this chapter, on p. 106. Because of their potential for adverse effects, the use of orally administered steroid treatment is only rather occasionally justified as a treatment for children with atopic eczema. Indeed, some paediatricians and dermatologists would consider that the use of steroids was never justified in atopic eczema, and would never prescribe them. This is not my view, because my own experience has satisfied me that they can be used with great benefit and as a general rule I have found that their potential for harm can be minimized.

It turns out that relatively high doses of these drugs are required to control severe atopic eczema. Short courses, of the type frequently used to treat bouts of asthma, only rarely result in anything more than very temporary benefit. A more sustained improvement in eczema generally requires continuous treatment over many months. It would not be appropriate to give very detailed information about dosage in this type of book, because different doctors will prefer different approaches. It is possible, however, to make some general comments. Initially, for successful relief of severe atopic eczema, relatively high dosage will be required. Perhaps the commonest reasons for oral steroids to fail is that the initial dose is too small. Very often, an attempt is made to reduce the dose too rapidly, with the result that the eczema quickly recurs. In order to plan treatment properly, one needs first to decide what is the aim of treatment. In practice, it is reasonable to aim to provide relief from the disease for long enough to allow a degree of natural improvement to occur that will allow discontinuation of steroid treatment, without recurrence of the eczema to its previous intolerable level.

Initially the most effective way to give these drugs is twice daily, in order to keep blood levels high most of the time. Once control has been achieved, the plan will be gradual reduction of dosage, and a change to once-daily administration. The advantage of once-daily administration is that the blood will not contain high steroid levels for much of the 24 hours, reducing the likelihood of unwanted effects and stimulating the body's adrenal glands to continue to function (see p. 107). However, once-daily administration may not be enough to allow adequate initial suppression of the eczema.

By reducing doses very slowly one can often maintain control of eczema very well, at dosage levels which would not have been sufficient to obtain that control initially. Ideally one would be hoping gradually to move from once-daily administration to alternate daily administration, as this further reduces the probability of unwanted effects, and would be even better at stimulating the adrenal glands. Unfortunately, alternate daily administration only very rarely proves adequate in practice. However, it may be possible to give smaller and larger doses on alternate days, which is better than giving the same total dose equally divided between the days. I have found that it may be necessary to continue this very gentle manipulation and reduction of dosage for a year or even longer before it is possible to discontinue the treatment altogether. However, in most cases it is eventually possible to stop the treatment without the eczema recurring to its previous level, so that less potent treatments can be used thereafter with satisfactory results.

There are several potential snags to oral steroid treatment with drugs like prednisolone or prednisone which require consideration. Growth is among the most important of these. These drugs usually, but by no means always, slow down children's growth. It is essential to realize, though, that this effect is decreasingly likely to occur as the dose is lowered, and that once treatment has been discontinued, catch-up can be anticipated, unless puberty has passed. For this reason, oral steroids should be avoided as far as possible during the period of rapid growth at puberty. This means that in general, treatment should not be discontinued by the age of 11 years in girls and 13 years in boys.

Another common problem with oral steroids is an increase in appetite and weight. Steroid hormones stimulate the appetite and lead to accelerated weight gain in this way. Steroids also alter the balance of water and salt in the body, leading to increased retention of fluid. They also cause a redistribution of fat around the body. The end result is a tendency to increased weight which is particularly noticeable on the face and trunk. Like the effect on growth, this is very variable, affecting some children more than others, and is really a cosmetic problem rather than a health problem. Therefore, it is likely to be seen as little of a problem in younger children, but may be a major one in a child approaching adolescence. Many children with very severe eczema have lost weight, and may look healthier when taking steroids by mouth than they did before, quite apart from the beneficial effect on the eczema. Again, these effects are temporary and will reverse as the dose is reduced. A longer-term problem can occasionally occur with the increased appetite, which may persist for a time after treatment is stopped. It can, however, always be brought back to normal with a little determination on everyone's part.

When oral steroids are given in high dosage, there is a tendency for the blood pressure to be increased. In practice, the doses used to treat eczema are rarely high enough to make this side-effect an anxiety. However, during the early days of treatment, when dosage is higher, it is worthwhile to have your child's blood pressure checked occasionally. Generally, if the blood pressure reading is substantially increased, all that will be necessary is a faster rate of dose reduction.

Very large doses of steroid may be required to treat certain serious diseases, and may result in an increased susceptibility to a variety of infections. However, this does not seem to be a problem with the lower dose levels that are appropriate for the treatment of atopic eczema. One precaution that is nevertheless advised in children taking steroids by mouth is the postponement of those immunizations in which live virus is given. Among the routine immunizations, these are MMR and polio.

Steroid hormone is essential to health, and a steady supply is required. Normally, the hormone is produced in the adrenal glands, and pumped into the bloodstream. Steroids taken by mouth have the effect of rapidly suppressing this production of steroid hormones by the adrenal glands. As a consequence, after a few weeks of having their activity suppressed, the adrenal glands may gradually lose their ability both to produce the quantities of steroid that would be required under normal circumstances, and to respond to the body's occasional need for a steeply increased production of steroids under emergency conditions.

The shutting down of the adrenal glands' daily steroid production means that the patient becomes dependent on the supply of steroids from outside. Therefore, if a day's treatment is missed, a steroid deficiency will develop; this will make the patient feel generally unwell and, if the deficiency is profound, it may lead to a fall in blood pressure, even to collapse and unconsciousness. This means that oral steroid treatment needs to be given on time, and that doses should never be missed. In practice, a day's treatment can generally be missed without ill-effect, but it is a good idea to add the missed dose to the next day's dose. If doses are being given twice daily and a dose is missed, add the missed dose to the next dose, whether it is taken the next day or later the same day.

Steroid hormone has particularly important functions when the body is subjected to severe physical stress, such as serious infection, acute profound illness such as heart attacks, physical trauma such as car accidents, or surgical procedures. Under normal circumstances, the adrenal glands are able to respond to such conditions by vastly increasing their steroid production. However, when their function is restrained by external steroid administration, they cannot respond effectively, and an acute steroid deficiency may be the result. Since the patient is already sick, the additional illness provoked by the steroid deficiency may not be recognized for what it is, and the consequences are potentially serious. In practice, this can be avoided by making sure that increased doses are given in appropriate situations. Usually it is sufficient to double the daily dose. This can be done if a patient has to have an operation, for example, or is ill. In the case of children, my feeling is that it is probably not generally necessary to give the double dose unless the child is ill enough to want to go to bed. As long as the child chooses to stay in bed, or at least to remain lying down, each dose should be doubled. Once the child is up again, return to the previous dose level.

In the case of older children, because of the danger that medical personnel might not know that steroid treatment was being taken in the event, say, of a road accident, it is a good idea to carry a card giving this information. It is probably even better to wear a Medic-Alert bracelet or pendant at all times (further information available from Medic-Alert Foundation, 12 Bridge Wharf, London N1 9UU, tel. 071 833 3034). In the case of children, you need to decide whether it is at all likely that your child will ever be out of the sight of a responsible adult. If your child is young, this is probably unlikely, and it is more important that you make certain that all adults who take responsibility for your child know about the steroid treatment so that they can inform ambulance staff, nurses, or doctors if it is ever necessary. Nevertheless, a Medic-Alert bracelet or pendant is a sensible back-up, while a card is unlikely to be helpful because it would be virtually impossible to make sure that small children carried it at all times.

As we considered earlier, the suppressive effect of steroid treatment on adrenal function can be reduced somewhat by giving doses as infrequently as possible. Thus, once-daily treatment is better than daily treatment, and alternate daily treatment would be better still. Alternating larger and smaller daily doses may be a good compromise if missing treatment altogether every second day is not possible. The suppressive effect reverses when steroid treatment is discontinued, though it may take several weeks for this reversal to be completed. It takes longer when steroid treatment has been given for very extended periods. It is generally considered wisest to discontinue steroid treatment rather gradually, particularly when the treatment has been given for longer than a week or two. This gradual withdrawal of treatment allows the adrenal glands to be brought back into action in a gentler way.

BECOTIDE®

'Becotide® is the trade name of a particular steroid called *beclomethasone dipropionate* (known for short as *BDP*), which appears to have rather unique properties. It was originally developed as an inhaled treatment for asthma. Some years ago, parents of children who had eczema as well as asthma pointed out that the treatment often improved their child's eczema, not just the asthma for which it was primarily prescribed. They particularly noticed this effect when dosage was fairly high. We subsequently took this observation further, and started to test BDP as a treatment for atopic eczema itself. We gave it by mouth rather than by inhalation, and found that it often worked well, if enough was given. We were particularly pleased to find that it allowed us to treat more successfully many of those children with extensive and relatively severe eczema for whom ordinary topical treatments alone had not alone been sufficient.

The special property of BDP that had interested us was that, once in the bloodstream, it is rapidly inactivated when it passes through the liver. This would make it much less likely to cause undesirable internal effects than more conventional steroids such as prednisolone. In practice, BDP does appear to provide many of the benefits of conventional oral steroid treatment without most of the disadvantages. In particular, there is no stimulation of appetite, no weight gain, and no increase in blood pressure. Its only unwanted effect is that growth is occasionally slowed. However, this is generally not a problem as the effect is not profound, and is reversible, so that full catch-up growth can occur later as the dose is lowered. The occurrence of this effect on growth in some children does mean, however, that oral BDP is best avoided at puberty, in order to ensure that adequate time is left for catch-up. We also prefer to avoid treating children who are already extremely small, for whatever reason. It follows that this treatment should not be given unless a child's doctors are prepared to measure growth regularly.

Becotide® is available in the form of capsules called *Rotacaps®*, which are designed to be placed in a small device called a *Rotahaler®*. The *Rotahaler®* punctures the capsule and allows the contents to be inhaled. However, for treatment of eczema, we ask parents to give the treatment by mouth. In this case, the *Rotacaps®* can be pulled apart and the contents dispersed in a small glass of water or orange juice. The treatment is tasteless and should be taken at least 30 minutes before anything else is eaten. Generally, a high dose is given at first. If the treatment is effective, the dose is very gradually lowered until the minimum effective dose is identified. Usually, growth is normal at this dose level, and treatment can then be continued for several months or even a year or two.

ACTH

As we considered earlier in this chapter (p. 109), ACTH is a hormone that stimulates the adrenal glands to produce increased amounts of steroid hormone. The advantage of giving ACTH (usually in the form of a brand called 'Synacthen®) is that it does not suppress the function of the adrenal cortex, though it does have all the other adverse effects of steroids, including suppression of growth. There are, however, three additional disadvantages. First, the response of each individual's adrenal glands is highly variable, so that it can be difficult to know how much effect one is going to get from any particular dose. Secondly, potentially dangerous allergic reactions can occur with this synthetic hormone, and thirdly it has to be given by injection. For these reasons, most doctors prefer to give oral steroids, but there are occasions when ACTH treatment may be more appropriate.

CYCLOSPORIN AND OTHER DRUGS THAT SUPPRESS THE IMMUNOLOGICAL SYSTEM

There is a variety of powerful drugs that appear to work by modifying the activity of the immunological system. Some of these drugs, such as 'azathioprine, have occasionally been used to treat extremely severe atopic eczema when all else has failed. The problem with these drugs is that they may cause a variety of side-effects which are generally considered too undesirable to justify their use except in very special circumstances. More recently it has become clear that a drug called 'cyclosporin, originally used to suppress transplant rejection reactions, can be extremely effective in treating atopic eczema. Cyclosporin is a very powerful drug with potentially serious side-effects, the most worrying of which is the kidney damage that some patients develop. Nevertheless, it seems likely that it will have an important role in treating very severe atopic eczema where all other treatments have failed. A considerable amount of research is currently being done on the use of this drug in severe atopic eczema. The aim of this research is to find the lowest dosage that will be helpful, and to identify more clearly the risks associated with its use at this dose level in patients with atopic eczema. The situation should, therefore, become clearer over the next few years. For the present, cyclosporin has not been used in eczematous children, but it is possible that it will turn out to be valuable for treating those very severely affected children for whom other safer treatments have failed for one reason or another.

LIGHT TREATMENTS (PHOTOTHERAPY)

The observation that atopic eczema often improves during holidays in sunnier climates, such as in the Mediterranean, led in northern countries to trials of treatment with artificial light. The therapeutic effect of light is due not to visible light, but to the invisible shorter wavelengths known as *ultraviolet light*. Strictly speaking, this should not be called 'light' at all because it is invisible to humans, and should more correctly be termed *ultraviolet radiation*, or *UVR* for short. UVR can itself be subdivided into two types, longer-wave UVR, known as *UVA*, and shorter-wave UVR, known as *UVB*.

UVB in sunlight is to a variable degree filtered out by the atmosphere. The higher the altitude, therefore, the greater the amount of UVB that reaches the ground. This is the type of UVR that causes sunburn, because its energy is absorbed in the very superficial parts of the skin, predominantly the epidermis. Sunburn in childhood seems to be the main predisposing factor in the development of a serious skin cancer known as malignant melanoma, and this type of cancer should therefore be regarded mainly as a consequence of excessive UVB exposure. UVB is also considered likely to be the main cause of the skin changes we associate with ageing, particularly the loss of elasticity which leads to coarsening of the features and wrinkling, and of skin cancers other than malignant melanoma. UVB is filtered out by many sunscreens to a variable degree that is measured by their *sun protection factor*, or *SPF*. The SPF is a measure of the additional exposure time that the sunscreen allows before burning occurs. Thus an SPF of 15 indicates that one would be able to spend 15 times as long in the sun without being burned wearing the sunscreen than one would have been able to without.

UVA is present in much greater amounts in natural sunlight than UVB. It is to a great degree responsible for the *tanning* response to sunlight. The role of UVA in ageing and in skin cancer provocation are not firmly established but there is considerable anxiety that it may play a part in both. The problem is that currently most sunscreens more or less exclusively filter out UVB, allowing the wearer to be exposed to vastly increased amounts of UVA before burning occurs. More recently, new sunscreens have been developed which also reduce the amount of UVA reaching the skin; these are more satisfactory.

UVB alone appears to be of rather little benefit as a treatment for atopic eczema. On the other hand, there have been reports, mainly from Scandinavian countries, of good results from treatment with combined UVA and UVB, resembling natural sunlight. To date, though, most effective of all has been a treatment known as *oral psoralen photochemotherapy* which comprises a combination of UVA and a drug given to the patient before exposure, called a psoralen. The psoralens are drugs obtained from plants, which effectively magnify by many times the effect of UVA exposure. The result is that a treatment effect can be obtained by relatively short exposure to UVA, measured in terms of minutes rather than the very long exposures that would otherwise be required. Although usually given by mouth, the psoralen can sometimes be painted on the skin as a lotion or by adding it to the bath-water. When the psoralen is given by mouth, the treatment is usually called *oral PUVA* for short, and when it is applied directly to the skin, it is called *topical PUVA*.

Oral PUVA has proved to be a highly effective way of treating atopic eczema. We have used it as a way of treating older children, particularly adolescents, who have eczema for many years and for whom other treatments have not proved effective. Although it can be a very effective treatment, it is difficult to administer because very expensive equipment (*Figure 55*) and highly trained staff are required, and because the patient has to travel for regular treatment sessions as often as two or three times weekly. Since it is not widely available, patients may have to travel long distances for treatment. Administering PUVA is also difficult, and requires great skill, because it can lead initially to an increase in irritation of the skin. As a result, treatment exposure needs to be very low at first, building up very gradually to the point at which it starts to work. How rapidly treatment exposure is increased depends on the irritability of an individual patient's skin, and requires careful judgement. Successful treatment needs experienced and determined staff, and determination on the part of the patient and family. The bonus is that oral PUVA offers a very good chance of a permanent improvement in a child's eczema, which sets it aside from other treatments that may produce good short-term results but which are less promising in terms of their ability to produce a lasting recovery.

The reason why this treatment is not offered more often is that there are concerns about possible long-term hazards. We are worried that its use may lead to an increased risk in skin cancer, and to the develop-

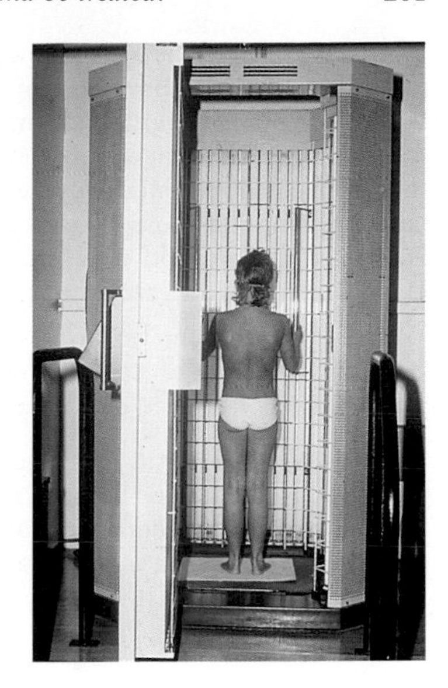

Figure 55 A PUVA cabinet, with the door open (reproduced with kind permission of St John's Dermatology Centre, St Thomas' Hospital, London).

ment of premature age changes in the skin. Whether these worries will turn out to be well-founded remains to be seen, but it may take as long as 50 years to decide whether the benefits of oral PUVA therapy warrant taking the risks that may be involved. For the present, I believe that it is a justifiable treatment in the most severely affected children, but only on the basis that a child gets one, or, at the most two, chances to obtain a permanent improvement in their eczema; if this doesn't occur the treatment should be abandoned while the total treatment exposure is still reasonable.

Topical PUVA offers the technical advantage over oral PUVA that the treatment can be more specifically localized to particular areas of skin. In practice, we have found that the applications themselves are irritating, and that the likelihood of a good response to treatment is less predictable. Furthermore, most of the children we treat in this way have very extensive eczema and are therefore more appropriately given treatment to their whole skin surface, at least initially. Later, if some areas require less treatment, this can easily be achieved by shielding those areas with clothing.

There are a number of precautions that must be taken by children receiving oral PUVA during treatment, such as wearing suitable eye protection; these will be explained to you in detail if this form of treatment is recommended by your child's dermatologist.

Recently there has been considerable interest in novel forms of phototherapy which have been made possible by the production of new UVR sources that allow high intensity treatment with rather specific wavelengths. These new lamps may enable us to identify and exploit those wavelengths which provide the principal benefit of phototherapy, while eliminating those that produce most of the undesirable effects. Advances in this field may ultimately mean that phototherapy will become a suitable treatment for a much larger proportion of children with atopic eczema.

SPECIAL SCHOOLING

Occasionally the best way of securing improvement in a very severely eczematous child is for them to obtain a placement at a special school for children with eczema, such as the Pilgrims School in Seaford. This option is discussed in more detail on p. 236.

CONCLUSIONS

You will appreciate from reading this chapter that there are many treatments for atopic eczema, and that no-one should regard the disease as untreatable. Many of these treatments include steroids, and I hope that I have managed to convince you that they have role in treating atopic eczema even in children. Steroids can make a substantial contribution to a successful treatment programme, and with adequate information they can be used with complete safety. However, they are not the only treatment for the disease, and excessive dependence on them will often lead to failure. It should be clear from reading this chapter that there is a wide choice of treatments that complement the use of steroids, and which allow successful treatment of some children without the use of steroids at all.

I have little doubt that successful control of a child's eczema leads to earlier natural resolution of the disease. This means that the more effectively you are able to treat the condition, the quicker it will disappear altogether. Keeping it quiet seems to give it a chance to get better naturally. I think of very active inflamed eczema as the skin in a bad mood, which it needs coaxing out of. Once in a better mood, you should find that gradually you are able to use less and less treatment to keep it this way until, with a little luck, you are able to discontinue treatment altogether.

9

Other problems in children with eczema

Certain physical characteristics, such as nail damage, enlarged lymph nodes, and extra eye-creases, are quite frequent in children with atopic eczema. These are direct consequences of the skin disorder and do not in themselves warrant any additional anxiety. However, children with atopic eczema are also liable to develop a number of other medical conditions which are worrisome to a variable degree. Some of these related disorders could be considered as complications of atopic eczema, urethritis for example, whereas others such as asthma are not direct consequences of eczema, and do occur in children without eczema; they are however much commoner in such children. A third group of problems, such as slowing of growth, have causes which are not well understood, and their relationship to the skin disease has not yet been defined.

NAIL DAMAGE

Nail problems are a common complication of atopic eczema. *Figure 56* is a diagram of the anatomy of the end of the finger, which I hope will make these problems easier to understand.

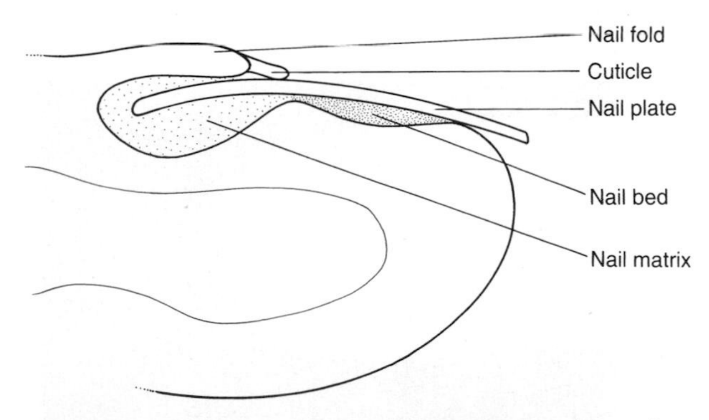

Figure 56 Diagram of the end of a finger, cut through lengthways from top to bottom, to show the nail.

The nail plate is made of the protein called *keratin*, exactly the same as the surface layer of the skin itself (p. 1), and the specialized part of the epidermis that produces it is known as the *nail matrix*. As it is formed, it is pushed outwards, stuck down on its underside by an adhesive tissue called the *nail bed*. The nail matrix is really a pocket of skin, in which nail is produced and from which it slides out on to the surface. The entrance to the pocket is sealed by the nail bed below and by the *cuticle* above. The cuticle is prone to damage if the fingers are affected by eczema. As a result, the bond between the cuticle and the nail plate is likely to be broken, and the route will then be open for bacteria and yeasts to get into the pocket. This is just the sort of warm, moist place they like, and the resulting infection in the top of the pocket, known as *paronychia*, will damage the upper surface of the nail, producing a series of transverse depressions and ridges in the nail. Where the infection is more intense, it may cause more severe damage to the surface of the nail.

Another type of problem can occur as a consequence of getting debris caught under the nail while scratching. If this debris contains bacteria and yeasts, infection can get a grip under the nail plate and this may be very painful. Such infections result in a small localized collection of pus under the free end of the nail (*Figure 57*). In cases where the infection is more severe, the nail plate may as a result lose its adherence to the nail bed, and occasionally nails are lost in this way. They will, however, always grow again. Avoiding this type of infection is another good reason for keeping the nails as short as possible to prevent anything becoming lodged underneath (see p. 124).

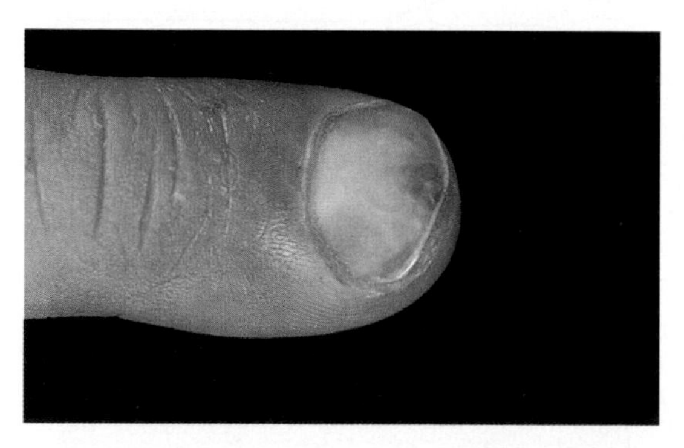

Figure 57 Infection under the free edge of the nail.

URETHRITIS

In children of either sex, if the genital area is affected by eczema, the opening of the passage through which urine passes (the *urethra*) may become inflamed. If this happens, there may be a burning sensation when urine is passed, and this can lead to anxieties about possible urinary infections such as *cystitis* (infection of the bladder). This problem is a very common one, and does not require treatment with antibiotics by mouth as urinary infections would. This area of the body should be treated with moisturizers and topical steroid applications no differently than any other area, and there is no evidence whatsoever to indicate that the use of steroids in the genital area is any more hazardous than it would be in other parts of the body. You should therefore not hesitate to apply steroid cream or ointment around the urethral opening if your child experiences this symptom. Of course, true urinary-tract infections can occur in children with eczema, just as in other children, but they are statistically much less likely to be the cause of this symptom. If there is genuine anxiety that your child does have a urinary infection, this can be reliably diagnosed by your doctor, by examination of a specimen of urine.

PALLOR

Most children with eczema have a curious pallid hue to their skin, generally most obvious on the face, which is often mistaken for anaemia. The pallor results from a somewhat reduced flow of blood through the skin in areas which are not affected by eczema. Where there is eczema, of course, the flow is more often increased. The significance of this altered pattern of blood flow is unknown, but it is believed that it is either due to change in the way that the blood vessels react to the various stimuli that affect their bore, or that it is a direct response of the blood vessels in areas relatively unaffected by eczema to substances released during the inflammatory process in affected areas and then circulated around the body in the blood.

LYMPH NODES

In Chapter 1, I described the function of lymph nodes (p. 3), and I mentioned them again in Chapter 7 (p. 74), when we were discussing skin infections. These bean-shaped structures are essentially filters for the tissue fluid (*lymph*), which passes through a series of lymph nodes on its way back into the bloodstream. The lymph nodes' principal task is to filter out bacteria, yeasts, and viruses from the lymph, as many of these would be very dangerous if they were to enter the bloodstream. In completely well children, these little organs are not very apparent, though they can be felt if one feels carefully at the back of the neck, in the armpits (*axillae*), and in the groins, for example. Under normal circumstances, one is also unaware of their activities. However, when they are having to work really hard they will generally become larger and may become tender. This is, as one might expect, commonly the case in the lymph nodes that drain lymph fluid from areas of eczematous skin.

Parents are often worried by the resulting tender lumps that they notice in their children's groins, armpits, and necks (see *Figure 34*). You should, however, regard these little lumps as, in a sense, an indication of the normal function of your child's protective mechanisms, and therefore as a sign of health. They are a witness to the efforts your child's body is making to prevent infection spreading from the skin into internal organs. Later on, when the skin improves, the lymph nodes will also settle down again, though this takes time. Doctors themselves sometimes express anxiety about enlarged lymph nodes, and this is especially likely to happen if a doctor does not have a great deal of experience of eczematous children. As a general rule, lymph nodes are not a worry in such children unless the enlargement is asymmetrical (much more apparent on one side of the body than the other) in the absence of an obvious asymmetrical skin infection, and so long as the nodes are smaller than a walnut.

ABDOMINAL SYMPTOMS

Small children with eczema seem to experience abdominal discomfort more frequently than other children, and they often also appear to have rather protuberant tummies (*Figure 58*). The reasons for these phenomena are unclear, and there is certainly nothing to suggest they have a serious cause. In practice, the great majority of these children grow out of these symptoms within a year or two without any special treatment being necessary. My own personal prejudice is to believe that there is a condition of the intestines that characteristically occurs in children with eczema, and which might almost be regarded as a sort of 'intestinal eczema'. This may to a greater or smaller degree be due to food allergies, but as with atopic eczema itself, it has proved very difficult to establish any definite relationship between these symptoms and what is eaten. I mentioned in Chapter 5 (p. 38) that there is definite evidence of some sort of rather mild intestinal abnormality in children with eczema; we know for example that there is a slightly increased 'leakiness' of the intestines which probably results from mild damage

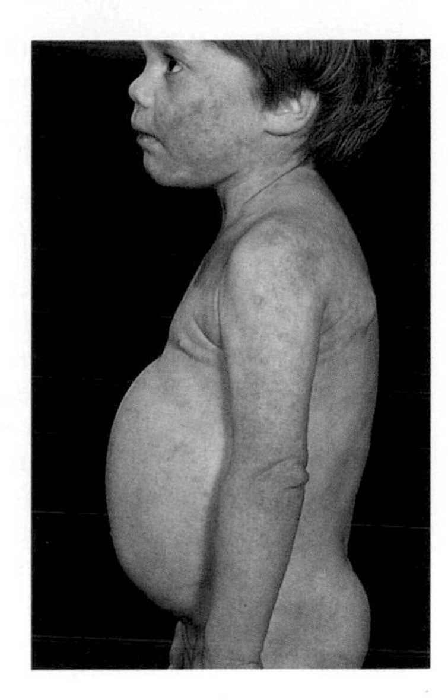

Figure 58 A typically protuberant abdomen in a three-year-old eczema.

to the intestinal lining. This condition seems, however, to be so mild that it probably never interferes with the absorption of nutrients from the food that the child eats. Just as in asthma, there may be spasmodic contraction of the muscle in the air tubes of the lungs, and it is possible that there may be similar spasmodic contractions in the muscle of the intestines in some children with eczema. These contractions could cause cramp-like pains, and might interfere with the normal passage of the intestinal contents from one end of the intestines to the other. By interfering with the passage of intestinal contents in this way, they could conceivably allow a degree of fermentation of sugars inside the intestines. Fermentation would be accompanied by the production of gas, which would cause more discomfort and bloating. Yet another contribution to abdominal discomfort could be made by inflamed nodes in the abdomen itself, where large numbers are situated, taking fluid up from the legs and from the abdominal organs into the chest.

You can see that it is difficult to say anything definite about the origins of the abdominal symptoms that one commonly observes in children with eczema. One can say, however, with a good deal of confidence that these symptoms very rarely if ever have any serious significance, and that they always seem to resolve with time and without the need for any special dietary manipulation or any particular treatment. Of course, it goes without saying that children with eczema are just as likely to have other types of abdominal disease as any other child. It is important therefore not to ignore symptoms that might indicate important diseases such as appendicitis.

ANAPHYLAXIS

Anaphylaxis is the name given to the most serious of all allergic reactions. Most typically, it follows injection of drugs, vaccines, allergen extracts (used for desensitization), or insect stings (i.e. injection of venom by wasps or bees, not 'bites' by blood-sucking insects such as mosquitoes). It results from the presence of IgE antibodies (see p. 8) to the relevant allergen. Although anaphylactic reactions are generally more common in atopic subjects they do not occur exclusively in such persons. Usually, but not invariably, the victim is already aware that an allergic reaction is a possibility because of previous though less severe reactions. Anaphylactic reactions vary somewhat in their severity. The most severe occur very rapidly, in the form of sudden collapse, with loss of consciousness, virtual disappearance of the pulse, and profuse sweating. More often, the onset is less abrupt, with giddiness, headache, a sensation of tightness in the chest, and sweating. These sensations are often accompanied by extensive wealing of the skin and swelling of the tissues of the face, particularly the eyelids and lips. Breathing may become difficult. Anaphylaxis in a child is extremely alarming for parents and also very dangerous. The more rapid the onset, the more dangerous it is. In hospitals with proper resuscitation equipment, it should be possible to prevent loss of life, but in other situations anaphylactic reactions may be quickly fatal.

Anaphylaxis is mentioned here because it is commoner in atopic individuals, and therefore commoner in children with eczema. If your child has symptoms of the type described after an injection or sting, even if they are fairly mild, you should take it as a warning that there is a risk of a more severe reaction the next time.

Although injections and stings are overall the commonest causes of anaphylaxis, this type of reaction can also follow ingestion of food. In this case, the reaction depends on the absorption of antigen into the blood from the mouth, throat, or intestines. The onset tends therefore to be slightly slower, but it may nevertheless happen surprisingly quickly. In my experience, anaphylactic reactions to injections and stings are in practice very rare in children with eczema. On the other hand, anaphylactic reactions to foods are not that infrequent. Though parents are often aware that their child is at risk of such a reaction to a food, they generally experience great difficulty getting others to take the risk seriously. Even medical people can be reluctant to recognize just how great the risk may be for a particular child, until it is too late. But parents will also have to convince friends, relatives, and those who look after their child at school, if they are to feel that the child is safe from being given the food which will cause the reaction.

The foods that most commonly provoke anaphylactic reactions in children in the UK are peanuts (strictly not a type of nut but a pulse) and eggs (generally egg-white), but other foods which less commonly cause the same type of reaction include fish, milk, and other nuts. The reactions may vary from exposure to exposure, and differ somewhat from the description given above of reactions to injections and stings. At their mildest, the skin may be almost the only affected part, with itching and blotchy redness of the skin, with or without wealing. There may be swelling of the face, usually most apparent in the eyelids or lips, and often this is oddly one-sided. In more profound reactions, the inside of the mouth may also be swollen, and tongue may become so large that it protrudes, and speech may become difficult. At the same time, breathing may become difficult, both because of swelling of the throat and because of asthma. There may also be abdominal pain and vomiting. In the most severe cases, consciousness may be lost. A fatal outcome is fortunately much rarer with food-provoked anaphylaxis, but it definitely can occasionally occur, and if your child has ever had a reaction of the type described, even mild, you should be aware that there is a risk of a more severe reaction on a future occasion.

If you have reason to believe that your child is at risk of an anaphylactic reaction to a food, you should take it seriously and make sure not to let anyone persuade you that there is no danger. Clearly you need to do all you can to make sure that your child is not given the relevant food by anyone. The precautions that will be appropriate will of course depend to a large degree on the age of the child. It will be relatively easier to protect your child during the pre-school years. Thereafter it becomes more difficult for a few years until the child is old enough to be able to look after him- or herself. During the early school years, if the risk appears significant, it may be justified to request a care assistant to supervise a child during breaks and at meal-times. It is also a good idea, even in younger children, to obtain a Medic-Alert bracelet or pendant (from Medic-Alert Foundation, 12 Bridge Wharf, London N1 9UU, tel. 071 833 3034), as this will warn those who deal with your child that the risk is serious, and will have the additional benefit of giving an important diagnostic aid if your child does have a reaction if you are not present.

If your child develops a reaction, you need medical assistance as soon as possible. The problem is that serious anaphylactic reactions occur so quickly that it may not be possible to get help while the reaction is at its height. It may therefore be worthwhile for you to have medication yourself to administer quickly should a reaction occur. The sooner treatment is given the more likely it is that you will be able to reduce the severity of the reaction. A large dose of an antihistamine such as 'Vallergan® helps, but is slow to work. The most effective treatment will be to inject 'adrenalin. If your child appears to be at substantial risk, and you feel you would be able to give an injection, it is worth asking your GP or consultant about this. There are available prefilled syringes containing adrenalin, or you can carry a glass vial containing adrenalin solution and an insulin syringe separately. You will need to be told the technique of subcutaneous injection and the appropriate dose (usually 0.01 ml per kilogram body weight, up to a maximum of 0.5 ml, of a solution containing 1 mg per ml; a single dose can be repeated five minutes later if necessary). If you are not keen to give an injection, or if you want a treatment your child can carry, an adrenalin inhaler can be a useful if somewhat less reliable alternative. The standard brand in the UK is 'Medihaler-Epi®, and the appropriate dose is 10–15 puffs.

Fortunately most, but not all, children grow out of this type of allergy after a few years.

ASTHMA

Asthma is a disorder of the tubes through which air travels in and out of the lungs. These tubes, or *airways*, have muscle in their walls that can contract and thereby decrease their bore. The bore of these airways can also be reduced by swelling of their lining as a result of inflammation, or by a build-up of secretions. Reduction of the bore of the airways will make it more difficult for air to pass through. Asthma is the result of periodic narrowing of these airways, in which all three mechanisms can play a part.

Just as in the case of eczema, there is a good deal of controversy concerning the precise cause of asthma, particularly whether it is always due to allergy. Whether or not allergy is an essential trigger, the immediate cause of asthma is an excessive sensitivity of the airways, resulting in intermittent narrowing which can be provoked by a wide variety of stimuli, including infections, irritants, exertion, and emotional stress. There are clearly close similarities with atopic eczema. Like atopic eczema, asthma is highly variable in severity. In many cases it is so mild that it is never recognized. In other cases it can be very disabling, and it can occasionally be dangerous. The best known manifestation of asthma is *wheezing*, a slightly musical note made mainly when breathing out, and often associated with shortness of breath. But wheezing occurs only during quite an intense attack of asthma. A much more frequent sign is coughing, often triggered by one of the factors I have mentioned. Coughing on exertion, coughing in bed, or coughing that persists for several days after a 'cold', are all particularly suspicious signs that suggest that your child may have asthma.

There is a general feeling among the experts that antigens which can become airborne, and which can therefore be inhaled, probably play a particularly important role in causing childhood asthma. The, particles which can be seen in their millions in shafts of sunlight mostly comprise fungal spores, pollen grains, fragments of shed skin, and house dust mite droppings. These particles will be inhaled, though only a minute proportion of them actually enter the lungs. Most of them stick to the moist walls of the nose, mouth, and the large airways leading down into the lungs. The lining of these airways is provided with millions of little hair-like projections (called *cilia*) that move in a coordinated way to keep a surface film of liquid moving upwards from the lungs. Any particles that are inhaled get trapped in this sticky film of liquid, known as *mucus*, which is then wafted upwards to the point where the main airpipe (the *trachea*) joins the back of the throat. The mucus then passes down the gullet (the *oesophagus*) into the stomach.

In recent years special attention has focused on house dust mite droppings as a cause of asthma. However, there is also evidence that foods may play a role in provoking asthma in some children, particularly younger ones. Nevertheless, as is the case with atopic eczema, allergy is not the whole answer.

Children with eczema are at relatively high risk of developing asthma. Like atopic eczema, asthma is a common disease and it probably affects at least 5 per cent of all people at some time in their lives. Asthma similarly tends to start early in life, before the age of three years in about half of all cases. Children who already have eczema appear to have at least a three-fold increased risk of developing asthma compared with other children, and perhaps as many as 40 per cent of them will need treatment for asthma at some stage. We have recently done research which showed that almost all children with eczema can be provoked into asthma under appropriate circumstances. In reality, it is probable that most eczematous children do experience a degree of asthma at some stage during childhood. However, it is also apparent that mild attacks of asthma are frequently not recognized.

A diagnosis of asthma need not disconcert you too much because most cases are mild from the start and remain so. It is only in the minority that real problems are likely. Like eczema, asthma shows a distinct tendency to improve with time, and most children with asthma will grow out of it after a few years. It can continue for many years, occasionally even into adult life. It may, like eczema, appear for the first time in adult life. Nevertheless, the most typical pattern is one of early onset and eventual more or less complete recovery.

You should seek medical advice if you suspect that your child has asthma. If it is confirmed, however, you should not assume that this means your child's life is going to be ruined by asthma, and for most it will never be more than an intermittent and minor problem.

One often hears that there is a 'see-saw' relationship between eczema and asthma, in other words, that where a child has both conditions, one will tend to be quiet while the other is active. In some cases, a child's asthma will tend to be little problem while the skin is bad, and the asthma will tend to be more troublesome when the skin is in good condition. I believe that such cases occur too frequently to discount the connection altogether, and I suspect that there is some truth in the story. It seems possible, in such children, that perhaps the inflamed eczematous skin is able to produce a substance that relaxes the airways. However, in most children the two conditions seem to behave in a more independent fashion.

It is also often claimed that asthma tends to make its first appearance when a child's eczema starts to get better. I am less convinced that there is any truth in this theory. It is true that the average age of onset of asthma tends to be a little later than that of eczema, so that in some cases, one will be improving as the other gets started, but overall they seem to behave in a pretty independent way.

NOSE PROBLEMS

Rhinitis is the medical term for inflammation of the internal lining (*mucosa*) of the nose. Rhinitis is remarkably common in children with eczema, and causes a variety of symptoms including runny nose, sneezing, blockage of the nose, and snoring. All these symptoms result from swelling and irritability of the mucosa. In children with atopic eczema, this inflammation appears to be caused predominantly by allergic reactions, and the condition is therefore called *allergic rhinitis*. Allergic rhinitis is generally due to the same airborne antigen particles we discussed in relation to asthma. In those cases in whom pollen is mainly responsible, symptoms will largely be restricted to the pollen season, causing what is popularly known as *hayfever*, or *seasonal* allergic rhinitis. If house dust mites, mould spores, or animal dander are responsible, symptoms will tend to occur more or less the whole year round, and the condition will be termed *perennial* allergic rhinitis.

The role of allergy is much clearer in allergic rhinitis than in either atopic eczema or asthma. There is strong evidence that mast cells (p. 10) play a leading part in allergic rhinitis. When an antigen particle, say a pollen grain, lands on the nasal mucosa, the wet surface causes the antigenic proteins within it to dissolve. These proteins are then absorbed through the mucosa, and come into contact with the mast cells in the underlying *submucosa*, the nasal equivalent of the skin's dermis. If the child is allergic to the antigen, IgE antibodies against it will have been made and will have become attached to the mast cells. Contact between the proteins and mast cells primed in this way will lead to the release from the mast cells of substances causing inflammation, such as histamine.

In fact, antigenic proteins can also probably reach these mast cells via the bloodstream, food proteins for example, and may also be able to cause rhinitis. Food-provoked allergic rhinitis is thought to occur most frequently in younger children.

Inflammation of the nasal mucosa causes it to exude fluid, which will cause a runny nose. The blood flow will be increased and may cause nose bleeds. Swelling of the mucosa often causes blockage of air flow, and affected children may only be able to breathe through the mouth, which can itself cause problems such as sore throats in the mornings. Nasal obstruction may also cause snoring. The lining of the nose extends into air spaces in the bones of the face, known as sinuses. Swelling of the lining of these sinuses can cause pain in the face or headaches. Still more problems may be caused by blockage of the *Eustachian tube*. This provides a channel through which air can pass from the nose to space inside each eardrum, and its function is to equalize pressure between the two sides of the eardrum. The clicking sensation in the ears when one lands in an aeroplane or descends a tall building in a lift is caused by a succession of closings and openings of this tube. Its lining is continuous with that of the nose and, if this lining is swollen, the tube may become blocked. This may cause ear problems, particularly infections (*otitis media*) which cause earache and hearing problems, both of which will need treatment in their own right.

Treatment is unnecessary for minor degrees of rhinitis, but if breathing through either nostril becomes difficult or if your child develops earache or any difficulty with hearing, you should consult your doctor. In the short term, the nasal blockage can be relieved by nose drops or sprays containing drugs that reduce blood flow, and thus reduce the swelling. These include such preparations as *Otrivine® and *Afrazine®. However, though effective when used for short periods, these drugs may aggravate the condition if used continuously. More effective long-term improvement may often be obtained by using the mast cell-stabilizing drug, sodium cromoglycate (see p. 179) (*Rynacrom® spray, drops, or cartridges) though regular treatment, at least four times daily, is required. Perhaps the most effective treatment of all is a steroid preparation, of which several are now available. These include 'Beconase®, 'Flixonase®, 'Syntaris® and 'Rhinocort®. Your doctor will be able to assess the necessity for any treatment and weigh up the pros and cons of the various available treatments.

I have a feeling that allergic rhinitis can make facial eczema worse. Allergic rhinitis tends to make the nose itchy, and an affected child may frequently rub the nose in an attempt to relieve this. The constant contact between the child's hand and the face may aggravate eczema in this area. Even more of a problem is caused by a runny nose, which may provoke eczema around the nostrils and above the upper lip. In these situations, treatment of the rhinitis may help the eczema.

EYE PROBLEMS

Most children with eczema have one or two extra creases in the skin below their eyes (*Figure 59*), which has the effect of making them look tired. These extra creases tend to be associated with rather a bluish tint to the skin around the eyes, which heightens the impression of tiredness. There are several theories about the cause of these changes. For example, it has been suggested that congestion of the nose and sinuses, due to allergy, leads to a build-up of fluid (*oedema*) in the upper part of the face, and that this is responsible. Against this idea is the fact that many eczematous children have this fold without any obvious nasal symptoms. However, perhaps the most likely explanation is that the extra creases are due to a mild degree of eczema. The skin around the eyes is rather loose compared to skin at most other sites and, as a consequence, it swells more easily. You can see this effect in everyday life; a mosquito bite will cause much more swelling if it occurs around the eyes than it would elsewhere. A little eczema of the eyelid area can therefore cause swelling of the skin even when it is otherwise unapparent. Rubbing of the eyes, an invariable result of eyelid eczema, will increase the swelling, and will of course lead in time to the development of some thickening of the skin at this site (*lichenification*—see p. 28), which would further add to the swollen appearance.

The blue hue to the skin below the eyes that one often sees in

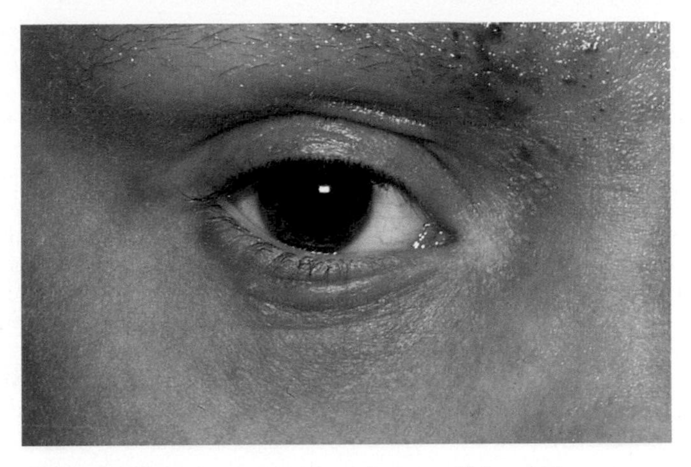

Figure 59 Extra eye creases in a six-year-old boy with eczema.

children with eczema is less easily explained, but may merely be related to the pallor that I shall discuss next.

The *conjunctiva* is a thin, almost invisible membrane covering the eye. Inflammation of this membrane is called *conjunctivitis*, which may have many causes including injuries, infections, and allergies. *Allergic conjunctivitis*, like allergic rhinitis, is common in children with atopic eczema, and may similarly be either seasonal, usually due to pollen allergy, or perennial, due to antigens that are present throughout the year. The seasonal variety is a characteristic component of hayfever, and is common in children with eczema. Conjunctivitis causes redness, watering of the eyes, and a rather gritty sensation. Like rhinitis, it can be treated with drugs causing decreased blood flow (such as *Otrivine-Antistin®) or by sodium cromoglycate ('Opticrom®). If very severe, steroid drops (for example, 'Eumovate®, 'Predsol®) are sometimes prescribed for a week or so, to get the situation under more rapid control, but such treatment should not be given without frequent supervision by a specialist eye doctor (*ophthalmologist*).

Eyes affected by allergic conjunctivitis are itchy and the child tends to rub them constantly. This rubbing can aggravate eczema around the eyes (*Figure 60*), so that treatment of the conjunctivitis will often help the skin as well as the eyes.

Vernal catarrh is the term given to a more intense form of inflammation of the conjunctiva, which also tends to be more severe during the spring and early summer. More intense discomfort is usual, and the

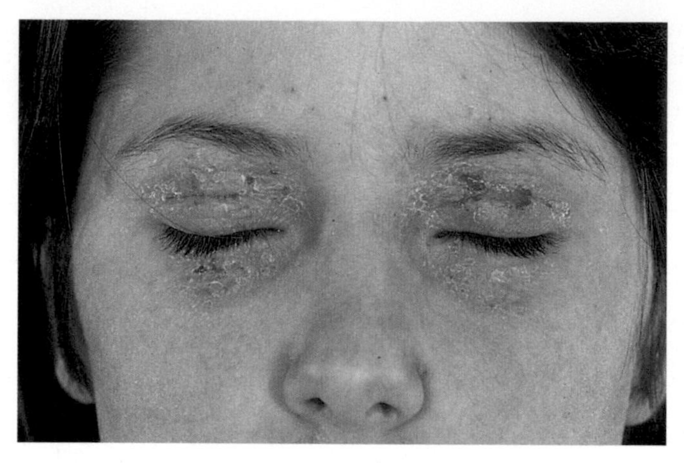

Figure 60 Periorbital eczema.

child may find bright light distressing (*photophobia*). Occasionally in this condition, the *cornea* (the transparent window at the front of the eye) may be damaged by friction with the inflamed underside of the upper eyelid; in this case the term *keratitis* or *keratoconjunctivitis* may be used. If your child has eye symptoms which are not controlled by regular use of 'Opticrom® you should request referral to an ophthalmologist for treatment. Vernal catarrh is likely to require steroid preparations to obtain initial control.

Keratoconus is a deformity of the cornea. Its cause is unknown, but it does seem to occur rather more often than one would expect in children with eczema. It is nevertheless a rare condition. Occasionally it causes some deterioration in vision in older children or adults with eczema, which can usually be overcome with glasses or contact lenses. Eye testing will pick up early signs of keratoconus.

Cataracts are opaque areas in the lens of the eye, which may appear for a variety of reasons. Children with eczema appear to have a slight risk of developing a type of cataract that is believed to be caused by substantial exposure to steroids. Usually this will mean that the child will have to have had orally administered steroids, but it is possible that it would occasionally occur as a result of absorption of excessive amounts of potent topically applied steroids. In practice, such cataracts are rare, and usually will not cause significant interference with vision. If they do interfere significantly with vision, the affected lens can be removed, but

I am keen to emphasize how rarely such action is necessary. They can be detected by eye tests, and it is worthwhile making sure that older children who have eczema have their eyes tested every year or two.

GROWTH

Children with atopic eczema are quite often smaller than they should be. Sometimes this is due to the fact that they also have asthma, which is known to interfere with growth in many cases. Sometimes a child's growth may be slowed by steroid treatment, either when it is given by mouth or when it has been applied directly to the skin; this issue was dealt with in some detail earlier (p. 108). However, there can be no question that atopic eczema itself can independently cause slowing of growth.

The mechanism by which a skin disease could affect growth remains a mystery. At one time, it was thought that disturbance of sleep might be the reason, since growth hormone is mainly secreted at night, but this hypothesis has recently lost favour. Growth disturbance definitely appears to be more likely in those children whose eczema is severe, and improvements in the condition, even when these are the results of treatment rather than natural resolution, can be expected to lead to an acceleration of growth.

If your child has eczema which is severe, it is a good idea to ask your GP or hospital consultant to take measurements of height and weight from time to time. Growth charts show the rates of growth of healthy children of different heights, and by taking measurements at appropriate intervals (perhaps six-monthly), you can tell whether your child's growth rate is normal.

A poor rate of growth indicates that something is wrong. I mentioned several particular reasons why growth may be slow in a child with eczema. A child with eczema is of course just as liable to other health problems as any child, but the danger that another problem may be overlooked is probably greater. Growth-hormone deficiency is an example of a type of growth problem that can occur in any child, and which one is keen not to miss because it is eminently treatable if detected in good time. A poor rate of growth in a child with eczema in whom causes other than the eczema itself can be reasonably confidently excluded indicates a need for more effective treatment. If good control of the skin condition is obtained, an acceleration in growth can be anticipated, and complete catch-up may be possible, on the one condition that the child is not by this time too old. Growth ceases at the end of puberty, at the age of about 15 in girls and 17 in boys.

10

Immunizations and eczema

There is widespread confusion about immunization and its possible dangers in children with atopic eczema. Even doctors may seem uncertain which immunizations are safe for such children, and which are not. The situation is in fact extremely clear, and any confusion is unwarranted. This is an opportunity to take a closer look at the whole subject of immunization, what it is, how it works, and what it is supposed to achieve, as well as considering it in the special context of the child with atopic eczema.

Two hundred years ago, it was well known to country folk that anyone who had a disease called cowpox would thereafter be protected from smallpox, at that time a common and dreaded infection. Cowpox, scientifically known as *vaccinia*, is a viral infection of cattle, but one which can also be contracted by humans if they touch an affected animal. The resulting lesion resembles a boil, and resolves within a week or two without needing any treatment. Edward Jenner (1749–1823) was the first to give people cowpox deliberately to protect them from smallpox. He did this by infecting them with the contents of a skin lesion from a dairy-maid, that is, by 'vaccinating' them. During the cowpox infection, the body starts to make antibodies which are soon able to eradicate the cowpox virus, and which will provide resistance (immunity) to further infections. Because the cowpox and smallpox viruses are in some details very similar, immunity to one confers immunity to the other.

Strictly therefore, the term *vaccination* refers specifically to smallpox immunization, but today the two terms are used more or less interchangeably. Immunization against infectious diseases can be achieved by the administration either of living micro-organisms (usually the same ones that cause the disease, but weakened or *attenuated* so that they will not cause full-blown disease), or micro-organisms that have been killed but which can nevertheless induce adequate antibody production. In some cases, immunization can be accomplished by administration of the outer and inactive coating (capsule) of a bacterium, as in the case of the new *Haemophilus influenzae* vaccine. Immunization for other diseases is accomplished by injection of small amounts of the bacterial toxin that normally causes the disease's effects, but which has been inactivated to make it harmless (*toxoid*).

The ultimate aim of immunization against a disease is frequently its eradication, an aim which has now been achieved at least in the case of smallpox. Of course, the more immediate aim of immunization is to protect the individual child who is being immunized.

However, there is another goal, an increasingly important one, but sadly one which is often forgotten. Let us consider measles as an example. Most children who develop measles are otherwise healthy. In a small number of these otherwise healthy children, measles will cause pneumonia, which is sometimes lethal. Others develop brain damage, which can cause permanent handicap. However, there are a number of children that are at greatly increased risk of dying or being permanently damaged by measles, because their general resistance to infection is compromised. An example of such a child would be one who is being treated for leukaemia, or one who has recently received a kidney or bone marrow transplant and is being given drugs to suppress the immune system (an *immunosuppressed* child). Such children cannot be immunized themselves because the attenuated virus in the vaccine remains potentially lethal to those who are immunosuppressed. Because of a much more successful measles immunization policy in the USA, measles is now rare in that country. In the UK, however, the number of children immunized against measles has until recently been much lower, and the disease remained relatively common. Apart from the complications that occurred as a result in otherwise healthy children, there were a much greater number of deaths from measles among our immunosuppressed patients than would occur in the USA. Fortunately, the advent of the newer MMR vaccine is rapidly improving the situation in the UK.

The protection of those who are especially at risk, but who cannot safely be immunized themselves, should be a major consideration for parents when they are deciding whether to immunize their own healthy children. Furthermore, should their own child ever develop a disease such as leukaemia, the child would then be at very high risk of death if exposed to measles, but would have been protected had they been immunized while still healthy.

The most frightening complication of an immunization in those with atopic eczema was *eczema vaccinatum*, a generalized and potentially lethal infection of eczematous skin by the vaccinia virus. This was a major hazard to those with atopic eczema until smallpox vaccination was discontinued. The disease could even be contracted from someone else who had recently had smallpox vaccination. The disappearance of this awful complication is of course an excellent example of what I was talking about earlier, i.e. the benefit to a highly at-risk minority of otherwise large-scale uptake of an immunization.

Table 6 Immunization schedule for children in the UK

Age	Immunization	Type
3–12 months	Diphtheria	Toxoid
	Tetanus	Toxoid
	Pertussis (whooping cough)	Killed bacteria
	Haemophilus influenzae	Bacterial capsule
	Polio	Live virus
6–8 weeks later	Diphtheria	Toxoid
	Tetanus	Toxoid
	Pertussis (whooping cough)	Killed bacteria
	Haemophilus influenzae	Bacterial capsule
	Polio	Live virus
4–6 months later	Diphtheria	Toxoid
	Tetanus	Toxoid
	Pertussis (whooping cough)	Killed bacteria
	Haemophilus influenzae	Bacterial capsule
	Polio	Live virus
12–15 months	Measles	Live virus°
	Mumps	Live virus° 'MMR'
	Rubella ('German measles')	Liver virus°
5–6 years	Diphtheria	Toxoid
	Tetanus	Toxoid
	Polio	Live virus
10–14 years	BCG	Live bacteria
	MMR (for children who have not previously had it)	Live virus
15–19 years	Tetanus	Toxoid
	Polio	Live virus

Let us now examine the currently recommended UK immunization schedule for children (*Table* 6), and then consider why anxieties may have been expressed in relation to children with atopic eczema.

Diphtheria and *tetanus* immunizations are with toxoid. The only risk is of an allergic reaction to the aluminium hydroxide that is added to slow down toxoid release, and thereby enhance the effectiveness of the immunization. In the past, the manufacturers used to provide rather vague instructions to doctors not to immunize 'allergic' individuals, but the relevant allergies were not specified. Doctors often interpreted this instruction as meaning that any child with any form of allergic disease was to be excluded from receiving immunization, and children with eczema or indeed, almost any rash, were generally not given it. There was in fact no justification for this policy, and the wording of the manufacturers' instructions has now been corrected to read: 'this vaccine should not be administered to a subject who has experienced a serious reaction to a previous dose of this vaccine or who is known to be allergic to any of its components'.

Pertussis (*whooping cough*) vaccine contains killed bacteria of the type that normally cause the disease. It is usually given in a combined vaccine with diphtheria and tetanus toxoid ('triple' vaccine). It is beyond the scope of this book to argue the pros and cons of pertussis immunization in general, but it is relevant to be aware that any risk from the vaccine is no greater in those with atopic eczema than in other children.

Haemophilus influenzae b is a very common bacterium which can cause a variety of serious illnesses, particularly meningitis. The vaccine consists of purified capsule (the protective outer coat of the bacterium), and is given by injection. Currently it is given separately from triple vaccine, but it is likely that in time it will be given together as a 'quadruple' vaccine. Adverse effects are rare and very mild, and there is no reason at all for not giving the vaccine to a child with eczema. Later 'booster' doses are not required as the bacterium is normally only a threat to those under the age of four years.

Poliomyelitis ('polio') vaccine contains live, attenuated polio virus, and is now given by mouth. The only children who should not be given this vaccine are those whose immune systems are suspect. This will include those very rare children with *serious* inborn defects of immunity (atopic eczema is definitely *not* among these), children with diseases that impair their resistance to infection (such as leukaemia), and children whose treatment severely compromises their resistance, particularly anticancer drugs. Children receiving steroids by mouth are excluded for the same reason, though there is probably little risk unless the dose is relatively high. Steroids applied to the skin, steroids such as 'Becotide® that are usually given by inhalation for asthma, or steroids such as 'Beconase® that are given nasally for hayfever, do not affect immunity and would not pose any risk. Therefore, live virus (or bacteria) vaccines are perfectly safe for all children with atopic eczema, with extraordinarily few exceptions. These exceptions are:

(1) any child known to have a *severe* defect of immunity, such as a failure of antibody production (*hypogammaglobulinaemia*)
(2) any child receiving prednisolone or a similar steroid by mouth, either for eczema or, more likely, for asthma, unless, under certain circumstances, the dose is a small one.

An alternative killed virus vaccine is available for the immunization of immunosuppressed children against polio.

Measles, mumps, and *rubella* vaccines also comprise live, attenuated virus, and so exactly the same set of rules apply as in the case of polio immunization. However, in the past, there was another problem, related to the fact that the virus used to make the vaccine was grown in eggs. This meant that the vaccines themselves used to contain traces of egg protein, and these could provoke reactions, sometimes severe, in anyone who was already allergic to egg. As readers will know, many children with atopic eczema are allergic to eggs, some exquisitely so, and such children were very much at risk from these vaccines. However, this type of vaccine has not been used for several years, because the virus is now grown in cultured chick embryo cells. Experience has confirmed that these newer vaccines are safe not just theoretically, but also in practice, even in those individuals who are liable to more serious allergic reactions to eggs.

The only vaccines that are still grown are *influenza* and *yellow fever* vaccine. These should not therefore be given to anyone allergic to egg, whether or not they have atopic eczema. Of course, not all those who have atopic eczema are also egg-allergic, but a proportion will be. If one of these immunizations is desired in an individual with atopic eczema, a skin test (prick test) should be used to exclude egg allergy, unless there is clear evidence of such an allergy from experience, in which case the test is unnecessary. These vaccines are *not* offered as a routine to children in the UK but, should immunization be recommended, the same rules apply as in the case of other live viruses such as polio.

BCG vaccine contains live, attenuated tuberculosis bacteria. At about 13 years of age, all children in this country are given a *tuberculin (Mantoux) test* to establish which are susceptible to tuberculosis. Those whose test is negative are offered BCG. In some parts of the country, BCG vaccination is undertaken routinely in early infancy. The same precautions apply to BCG as to the other live vaccines—it should not be given to anyone who is immunosuppressed or is taking oral steroids. In addition, the manufacturer's instructions accompanying the vaccine do state that it 'should not be given to patients suffering from generalized eczema'. However, there is no real evidence that BCG is ever dangerous to those who have atopic eczema. The theoretical hazard is that the bacteria comprising the vaccine will multiply in the eczematous skin, and therefore cause a more extensive infection than the usually very circumscribed one. Such a complication of BCG immunization has never actually been reported, either to the manufacturers or in the medical literature, and it seems highly improbable that it ever occurs in practice, as many thousands of children with atopic eczema have now received this immunization despite the warning.

In families where the risk of a child developing atopic eczema seems high, parents often worry that immunizations may cause or at least provoke the onset of eczema. This fear has often persuaded them to postpone or avoid these immunizations. In fact, there is very little evidence that such a danger exists, and it is strongly advised that immunizations be given on schedule.

CONCLUSIONS

The situation is really very simple. *None of the vaccinations routinely offered to children in the UK is any more hazardous to children with atopic eczema than to other children.* Neither is there any strong evidence that these immunizations cause atopic eczema. The routine immunizations should all be given at the recommended time. Influenza and yellow fever vaccination should currently be avoided by anyone who is egg-allergic, but these vaccines are not offered routinely. Immunizations with live viruses (polio, MMR, rubella, influenza, hepatitis A, and yellow fever) and bacteria (BCG) should in general be deferred in children taking oral steroids.

11

Living with an eczematous child

Children who have eczema can be quick to appreciate its great manipulative value, and many do not hesitate to exploit this to the full. Children with eczema soon become aware that their eczema can be used to secure special treatment, so long as a failure to secure this special treatment is marked by a fearsome bout of scratching. The drawing of blood is a powerful weapon. The more effectively the technique works, the more it becomes reinforced and established as a pattern of everyday behaviour. Children learn to use a handicap like eczema in this way quite naturally, instinctively, and, at least initially, subconsciously. If it proves successful, the manipulation may later on become a more conscious one, and this is when it becomes undesirable and hazardous to the child's development as a normal member of society.

What can be done about this? Being aware of the problems is half the battle. Do not let the pattern of events develop in which a bout of scratching will always end up with your child getting exactly what he or she wants. Tantrums which take the form of scratching have to be ignored just like any other kind. Once your child realizes that this sort of thing won't work, he or she will stop doing it.

Children with eczema need not be scarred by the disease itself, either physically or mentally. They may, however, suffer lasting harm as a result of having been treated as 'different' from other children, even though this may have been done with the best possible motives. It is for this reason vital that, as far as possible, you treat your eczematous child no differently from the way you would treat any other child.

Parents often imagine how they would feel if they had the eczema instead of their child. Fortunately, the emotional effect of severe eczema on a child is not the same as it would be on an adult. Small children live from day to day, even from minute to minute. Luckily they do not look ahead sufficiently to become disheartened, and quickly recover from bad times. However dreadful their eczema has been, they generally will quickly forget and get on with their lives. Many parents suffer more mental anguish than their eczematous child. This is a great pity, because it can impair parents' ability to cope with their child's needs.

The great majority of children with eczema will eventually recover physically. However, it is to a great extent up to parents to make sure that when this happens they will be no different from other people. Most important of all is not to be over-protective. Wherever possible, let your child do the things that other children do, even if it does make the skin a little worse. It is of course perfectly reasonable to try to interest your child preferentially in things that do not make the eczema worse, but if your child desperately wants to ride horses or go swimming, activities which can have an aggravating effect, don't stop him or her from doing so. Don't ask if your child can be excused PE or games at school, and don't stop your child going to birthday parties because of special diets. A single afternoon's deviation from a diet is unlikely to do as much harm in the long term as preventing a child from joining in the fun. This is the point, not to pay for short-term gains by long-term losses.

Children with eczema often spend too little of their time in the company of other children, and correspondingly too much in the company of adults. Often they are expected to behave in ways that would be more appropriate for an older child, and the result is a child who grows up too quickly. While such children are often found charming by adults, the phenomenon is telling us that the child is missing out on something very important to normal development—childhood itself. Do all you can to ensure that your child spends plenty of time in the company of other children of his or her own age.

Tension often builds up in families where a child has eczema. We have already considered the tension that can be generated by the child's attempts, conscious or subconscious, to get his or her own way. Your other children will resent any additional attention given to a brother or sister with eczema, and this is another reason why you must make quite sure you are not manipulated. Some extra attention is unavoidable, because it takes time to do treatments, and because diets may mean special food. It is important, however, to be aware of the jealousies that such 'special' treatment may arouse. If you are careful to compensate for it, it need never be a problem. Husbands and wives may envy the attention given to an eczematous child, in just the same way. Arguments about the child's treatment are common, and are often symptomatic of this kind of resentment.

Such arguments are liable in their turn to generate additional tension in the family. Occasionally, such disagreements are actually capitalized on by a manipulative child. Children can appear to be unable to recognize that their own divisive behaviour is threatening family harmony. Indeed the increasing disharmony will itself tend to aggravate the very behavioural traits that caused it in the first place.

It is very important indeed never to let your child become aware of the worry you feel over his or her skin disease. The same applies to other members of the family, especially grandparents, aunts, and uncles. Such relatives, and your own friends, may become a source of very unwelcome advice. You need to handle this type of 'help' very firmly. Discussion of eczema should be restricted, at least in the presence of the affected child. While it is normal to talk about a child's eczema, this topic should never be allowed to dominate conversation in the affected child's hearing and you must make it clear to relatives and to friends that this is how you want it.

If you complain to relatives and friends about your doctors, you are asking for trouble; you will have given them 'carte blanche' to give their own advice instead. Be confident in your approach. Listen to what others have to say, but make it clear that you will make the decisions you believe to be in your child's best interests. Do not take every wacky suggestion that they may make too seriously, but always express your appreciation of the fact that they care. You may feel that their care could be expressed in ways that would be more helpful to you and to your child. For example, it may be a great relief to you (and to your child!) to have him looked after by someone else from time to time. This will definitely help to reduce your own feelings of tension, and will give the child a break from exposure to this tension. Ask whoever is willing to take this responsibility to do exactly those eczema treatments you do yourself, and not to use the opportunity to undertake therapeutic experiments in which your child is the 'guinea-pig'!

Do all you can to avoid feeling that you are the only one who could possibly cope with your child; this is important in the case of any child, but even more so in a child with any kind of problem. If your child's eczema always looks much better after a few days with someone else, ask yourself why this might be? Did they merely do the treatments more carefully or more regularly, or did your child benefit from a less stressful environment?

People outside the family constitute another potential source of tension. Some people will not recognize what is wrong with your child and may think that the child has an infection; others will be convinced that the child has been burned or physically harmed. The boldest will say so openly. This is one problem that parents almost always encounter, sooner or later. Do your best not to be upset by it, and avoid becoming victim to a type of paranoia in which you interpret any look given by a stranger as hostile or accusing. This may mean that you become hesitant to take your child out into public places because you dread people's curiosity, and because you fear the comments that they may make. You may even be afraid of reacting violently yourself. However, this isn't a good idea. You should never fear taking your child anywhere. Remember that he will quickly pick up the idea that you think his appearance is so offensive that you want to hide him away, and he will respond by becoming ashamed himself. Also remember that curiosity is a very natural human trait, and one that is present in every one of us. There is nothing wrong with curiosity itself; it is just that it goes too far in some people and in these individuals, it may be combined with tactlessness. The best thing is to respond quickly to sustained curiosity by explaining that your child has eczema, and telling the person whatever seems appropriate about the problem. Your confidence will be the best protection against undesirable comments, and will teach your child that there is nothing to be feared. Encourage your child to be open and factual with other children. Whenever the opportunity arises, strike a blow for eczema sufferers everywhere by gently explaining that eczema is not the result of lack of hygiene, nor is it contagious. Don't hide your child away; more good is done by letting people get used to the idea that he or she has eczema.

The next problem with outsiders is that, like friends and relatives, they will often have views on how eczema should be dealt with. There are hundreds of theories about the cause of the disease, and lots of busybodies keen to tell you of miraculous cures. When people offer their theories, listen politely. If you don't want them to, try to avoid eczema as a subject of conversation, and especially try not to appear too desperate. Again, do not complain to outsiders about your doctors; this won't help!

DEALING WITH SCRATCHING

I am often asked by parents what they should do about their child scratching. This is a very tricky problem. An answer to the scratching problem would be an answer to the whole problem of eczema, and the two are inseparable. Parents' distress at seeing their child tearing himself to pieces naturally leads them to try to intervene. The problem is that you know that they may not stop scratching until the itching has been replaced by pain, by which time damage will have been caused that will take days to heal. Older children recognize the link between uncontrolled scratching and worsening of the eczema, but younger ones do not. This is very harrowing for parents. Unfortunately, intervention on your part may seriously aggravate matters. Never shout at your child to stop doing what is partly an involuntary reaction to the intense itching. If you do, you will only heighten the tension and further intensify the irritation. Your child will interpret the shouting as a sign that you do not love him or her. The same is true of any physical attempt to prevent scratching. Never punish a child for scratching.

Having said what you shouldn't do, what should you do? All you really can do is, firstly, to make every effort to distract the child's attention away from the skin and, secondly, to reduce the amount of damage done as a result of the scratching. Distracting the child is more difficult than shouting, and more time-consuming, but it is the right way. Do anything that will focus attention on something other than the skin. This will require a different approach in every child. It may mean going out of your home, or wherever you are; it may mean playing a game. If simple distractions fail, it may help to pop the child in a bath; this can often provide surprisingly effective relief. An alternative is to apply generous quantities of a suitable moisturizer (p. 101). This will at least provide a layer of protective lubrication to reduce the damage done by any scratching. If all this fails to relieve the irritation, or if you feel that the scratching is being used to gain control over you, you may find it best to let your child scratch, but with the least possible destructive effect. Make sure always that the fingernails are as short as possible and free of jagged edges (p. 203). Encourage your child to apply a moisturizer while scratching. Try to promote rubbing rather than scratching. It is sometimes a good idea to do the rubbing or scratching yourself, but gently. Some parents find that they can make deals with their child; the child is allowed to scratch without interference for a fixed length of time, say two or three minutes, on condition that the scratching stops at the end of that time.

SLEEP PROBLEMS

For many parents of a child with eczema, nights are pure hell. Eczematous children almost invariably have problems with sleep, which often become the most difficult aspect of their care. There are in fact several different types of problem, all of which tend to result in disturbance of their parents' sleep. There may be difficulty getting off to sleep in the first place, there may be frequent waking during the night, requiring parents to provide comfort, and there may be dreadful difficulty getting a child into action when the morning finally arrives. The most common one, which few parents of a child with eczema completely avoid, is their child waking during the night, starting to scratch, and needing an awful lot of attention and effort on the parents' part to get back to sleep. This may happen once during the night, or many times, and parents can go for nights in succession without satisfactory sleep themselves. For many parents, this is the worst of all the bad things that having a child with eczema can mean.

Sleep comprises a constant cycle of lighter sleep, technically called *rapid-eye-movement sleep* (*REM sleep*), because of the rapid and irregular eye movements that characterize it, and a more relaxed sleep called *non-rapid-eye-movement sleep* (*NREM sleep*). In a night of normal sleep, each cycle comprises about 75 minutes of NREM sleep followed by about 15 minutes of REM sleep. Adults enjoy four or five cycles in a normal night. Dreaming seems to occur predominantly during the REM phases, which become longer as the morning approaches. Paradoxically, one is not at one's most wakeful during REM sleep, but during the final part of the NREM phase just before the switch to the next REM phase. I suspect that what happens is that a child with eczema may reach sufficient consciousness during this brief more wakeful period between cycles to become aware of itchiness. A few scratching movements at this point may then be enough to animate the child a little more, then a little more again, until he or she is fully awake. Interestingly, when the child gets back to sleep again, another cycle is started, and little overall disturbance occurs. However, when the child wakes its parents, they are likely to be awoken during a different phase of the cycle, which is more disturbing, and results in greater deprivation of the beneficial effects of sleep.

Parents will usually go to their child when he stirs, in the hope that they will be able to settle him before he becomes too irreversibly awake, and in the hope of averting too much skin damage due to uncontrolled scratching at such times. While these are very laudable motives, it is worth bearing in mind that children like to have their parents with them at night, and may get to like the whole process so much that they take to calling or going along to their parents' bedroom as a matter of habit, saying that they are itchy, but really just seeking fun as much as genuinely needing comforting. Having said this, if your child really does need attention in the night, a few tips may be helpful.

Babies with eczema often wake because they are hungry or thirsty, just like any other baby. The first thing you should do therefore is to try providing a feed or a drink. Babies and older children can often be soothed very effectively by gently scratching, rubbing, stroking, or massaging their backs. If you can accompany this with a lullaby, all the better! It is often a good idea to establish a little ritual which you go through on each occasion that you go to your child, the end of which the child recognizes, allowing you to return to bed without problems when this point is reached. The ritual might for example consist of changing a nappy, or using the potty or toilet, followed by a drink, an application of moisturizer, and ending with a few minutes of gentle stroking.

Many parents allow or even encourage their eczematous children to share the bed with them. Often this is done in the hope that a parent will be alerted to a bout of scratching in the night before this becomes too well established. Other parents find that the child simply sleeps better in the close proximity of a parent, or that they can coax the child back to sleep more successfully if they do not have to leave the child immediately. Other parents simply find that it is less disturbing to their own sleep to manage things this way, particularly if one parent goes to sleep elsewhere. Every family has to find the system that works best for them. You definitely shouldn't feel guilty or unusual if you have your child in bed with you and not your spouse; this is very common in households with an eczematous child. However, if you do share a bed with your child, you must remember that they do not like to get hot. Your own body heat may be a problem in this respect, so the bed will need to be big enough that you don't have to be too close. The bedroom may have to be cooler than you would normally yourself like it, and the bedding may have to be lighter than you would normally want for yourself.

Another factor you should consider if your child is sleeping very lightly is whether the child can breathe properly through his nose. Nasal obstruction due to allergic rhinitis is common in children with eczema (see p. 212), and may interfere with normal sleep patterns. Suspect that this may be relevant if your child snores or breathes very noisily at night. Treatment of rhinitis in such children can be extraordinarily effective in improving sleep. Of course, allergic rhinitis is not the only cause of nasal obstruction; your child could have adenoid enlargement, for example, which has nothing to do with eczema.

HOLIDAYS

Whether, where, and when to go on holidays—these are important questions for the parents of children with eczema. As with so many questions about eczema, there are no simple answers. Nevertheless, a degree of generalization is possible.

More often than not, holidays are quite beneficial for children with eczema. The most beneficial of all seem to be holidays in sunnier, less damp, climates. The Mediterranean is most reliable of all, but other sunny climates also usually have a positive effect, including the southern USA, Africa, the Middle East, Asia, and the Far East. Atopic eczema is in any case much less common and generally less severe in these parts of the world (though it is becoming more of a problem whenever air conditioning is in use for much of the year). The improvement experienced in the eczematous visitor to these parts of the world does not seem to be a result of sun exposure, nor of swimming in the sea, though these factors certainly may contribute. There seems to be something generally soothing about the climate as a whole, but exactly what this may be remains unclear. A reduction in exposure to important non-food allergens may be another factor. For example, exposure to house dust mite droppings is likely to be very low in a hotel in Spain, in which floors will probably be made of stone or marble, and where the child is going to spend much of the day out of doors in any case. However, changes in allergen contact are probably not the main factor either. One of the reasons I say this is because, at least in the case of food allergens, parents often find that their child can eat with impunity things that would definitely make the eczema worse at home. This suggests that the effect of the holiday is to make them more tolerant of the factors that would usually lead to aggravation of the eczema, which would include both food allergens and non-food allergens.

The main threat to eczema in warmer climates is the increased risk of skin infection. Heat and humidity both encourage the proliferation of bacteria in eczematous skin; the hotter and more humid it is, the greater the risk of infection. The infection risk is greatest during the first few days, before the eczema improves. It is important to be aware that risk can be significant if the weather is really hot and humid. For this reason, I always advise parents to avoid parts of the world where the climate is *very* hot, and to avoid the hottest times of the year in some other areas, such as July and August in the Mediterranean.

If you are on holiday abroad and your child's eczema does rapidly deteriorate, the odds are that a bacterial infection of the skin is the reason, and you should seek appropriate medical advice and treatment.

One other factor worth bearing in mind when planning a holiday is that long journeys in cars, trains, and especially in aeroplanes can aggravate eczema in the short term. Boredom is a factor, but the intense dryness of the air in aeroplanes is probably the most important one, together with the interference with sleep patterns caused by time zone changes.

YOUR OWN SANITY

Having a child with eczema can lead you to feel desperate. Few adults who have not had the experience can imagine what it can be like, and you will feel that no-one understands. This can lead to a sense of isolation. Clearly, some will be more fortunate than others in having understanding partners, relatives, or friends. Some doctors will be more sympathetic than others. You may find it very helpful to contact the National Eczema Society (tel. 071 388 4097) (see p. 248), who should be able to put you in touch with someone in your area who knows about eczema and the problems it causes. Being able to talk about your difficulties to someone who understands is half the battle.

The demands of an eczematous child can push a wedge between father and mother, particularly if they are not even able to enjoy one another's company at night. The problems you are experiencing with your eczematous child will pass, and it will be important to ensure that you still have a relationship to share. So, do all you can to keep that relationship going. Try to find someone who will look after the children for a few hours each week when the two of you can go out somewhere together.

FINANCIAL PROBLEMS

Caring for a child with severe eczema can be expensive as well as hard work. You may be entitled to claim for a Disability Living Allowance, a benefit which replaces the Attendance and Mobility Allowances. To qualify for the care component of the new benefit, you need to show that the level of care required by your child is substantially greater than would be required for a perfectly healthy child. This is certainly likely to be the case if your child is severely affected. The mobility component is less likely to be relevant, as most children with eczema are reasonably mobile despite their illness. To make a claim for the Disability Living Allowance, go to a Post Office and pick up a leaflet about it, telling you how to obtain an application form.

A number of charities may be able to help with specific expenses in the case of children who are substantially disabled. A good example is The Family Fund (PO Box 50, York YO1 2ZX), to whom you can apply directly for help with any particular item you consider would be helpful to your child. These might include clothing, shoes, bedding, a washing machine, tumble drier, heating, even an outing or a holiday where appropriate.

If you are finding care of your eczematous child a financial problem, carefully consider making an application to one of these sources of assistance.

12

Education and career planning

It has often been suggested that children with eczema are of above-average intelligence. However, careful studies have shown that this is a myth, and it is more likely that the impression of maturity given by many eczematous children reflects their tendency to spend more time with adults than would be the case in most children. As a result of this excessive association, they often seem very much at ease in the company of adults, and tend to take on adult ways of communication, a phenomenon that has sometimes been called the 'cocktail party syndrome'. This habit tends to be perpetuated by the fact that adults find it acceptable and are often clearly impressed, saying how mature or grown-up such a child is.

In fact, children with eczema are on average no more or less intelligent than other children. Intelligence is a measure of potential and, in a sense, what is more important is actual achievement. So, do eczematous children actually perform as well academically as their counterparts? The answer to this is that overall they almost certainly do not. Their educational progress is hampered in a number of ways. The most obvious is their frequently irregular attendance at school. In addition, they are likely to find it difficult to maintain concentration in the face of any degree of skin irritation. Attentiveness can also be impaired by the soporific effect of antihistamines (p. 149). I also have the impression that some parents actually try to shield their eczematous children from the academic pressures of school, in the belief that these pressures represent an unfair additional burden on them. But, as I suggested in the previous chapter, any action that tends to isolate children with eczema is likely to do long-term harm. As in all other facets of your child's life, in educational matters it is important to try to provide an environment that is as normal as possible.

The education problem is, of course, more or less restricted to children with more severe eczema; the worse the eczema, the greater the possibility of a problem. Paradoxically, education is of particular importance to those with severe eczema, because a wide variety of possible careers are going to be unsuitable for them. Any job which involves contact of the skin with irritants is best avoided, and this limitation should be borne in mind throughout the child's education.

A few principles may help in the successful education of eczematous children, though they will not all be appropriate in every case. I am sure that any education professional reading this book would be able to add to these.

It seems to me that there is often much to be gained from eczematous children starting school early, wherever possible, by getting them into nursery school. This may help to give them a head start to compensate for absences that may occur later. But, more importantly, I think they often benefit from the social experience of school, and from the periodic physical separation from their mothers that school entails. This can provide the basis for a less intense and a more independent relationship between mother and child, something which is mutually beneficial. The greater activity that goes with school turns the child's attention outwards, helping to prevent development of the introverted personality that so often goes with eczema. School activities also tire a child, and may help them to develop a better pattern of sleep.

When your child is about to start at school, it is a good idea to speak to his or her prospective teacher. Tell the teacher that your child has eczema, and explain the kinds of treatment being used. If the teacher knows little about eczema, a little general information may be welcomed, and you should particularly lay emphasis on the fact that the disease is not hazardous to others. Other pupils, parents, or staff may later ask the teacher what is wrong with your child, and it helps if clear answers can be given them from the start. Make sure to have a word with the school nurse when there is one. Ask her if she would be prepared to apply creams and ointments if necessary. If she will do this, it is a good plan to provide her with a supply of whatever moisturizer you prefer. This helps her by giving her something she can do for your child when and if the need arises. If your child is on any special form of diet, it is obviously important to check that the school can cope with this. If not, you may need to provide packed lunches. Only take a child home for lunch if it is absolutely essential as it will result in them being excluded from normal playground activity.

Check that your child will be seated in the coolest part of the classroom, as far away as possible from radiators, and, in summer, away from direct sunlight through a window. It is a good idea for the child to be as close to the front of the class as possible, to try to maximize attention to what the teacher is doing and saying.

Do your best to make sure that furry or feathered animals are not kept in the classroom (see p. 52).

As often as you can, take the opportunity to tell other children's parents what is wrong with your child, as this will help to prevent those children being told to keep away. If your child seems upset at the end of a school day, you could consider asking if he has been taunted about his skin—this can help by bringing the problem into the open. However, don't forget to ask about the things that went well at school as well as those that didn't. If you discover that your child is being abused, verbally or any other way, because of the eczema, let the teacher or headteacher know. Remember that it is an important part of the education of normal children that they come to terms with those less fortunate than themselves.

Sometimes it is suggested to parents that their eczematous child might do better at some type or other of special school. It seems to me that such recommendations often reflect the inadequacies of the school and its staff, rather than any real need for change of school from the child's point of view. As in all aspects of your child's life, as normal as possible an approach is best. If you want help with this kind of problem, do not hesitate to approach your doctor and suggest that you discuss the situation with a social worker. Occasionally a change of school is a good idea, but this is probably the exception. Boarding schools, if you can afford to send your child to one, can sometimes be a great help, with their greater emphasis on outdoor activities and self-reliance. However, if you do consider sending your child to a boarding school, check that the dormitories are kept dust free and if possible that they are carpetless, and that your child is not likely to have to sleep in a bottom bunk-bed.

The physical education side of school activity can be a special problem. Your child may find undressing embarrassing, and this requires sympathy on everybody's part. The sweating induced by hard physical exertion can lead to quite marked irritation of the skin and, where there is a choice, activities that lead to the least irritation should obviously be preferred. Nevertheless, as far as possible, eczematous children should take part in physical education and games to avoid being different, even if a degree of aggravation of their eczema is the price. Do think hard before suggesting that your child be exempted altogether from sports.

Absences from school can become a problem. As far as possible, you should avoid keeping your child away from school, mainly because it singles him out as different. Very occasionally, frequent absences are inevitable, and such a child will need extra help. This is something you must discuss with your child's teacher. Home tutoring is sometimes useful for children who have fallen behind, and in every case that I recall, this has been very successful in enabling a child to catch up and re-enter the normal education system in due course.

In summary, do all you can to get your child as good an education as possible. The emphasis should be on normal treatment at school. Your main task is quietly to try to prevent problems before they arise, by talking to teachers, school nurses, and other parents. I am constantly heartened by how well severely eczematous children can do, in spite of all their problems. Understanding on the part of teachers and fellow pupils will help them immeasurably.

SPECIAL SCHOOL FOR CHILDREN WITH ECZEMA

As far as I am aware, there is only one school in the UK which caters specifically for children with eczema. This is the Pilgrim's School in Seaford on the English south coast. A placement at this excellent school can be a life-saver for a child with severe eczema and for their families.

Children will generally be placed at the school by local education authorities, who pay the fees and who will usually only contemplate such a measure if the situation is a desperate one, particularly if there is clear evidence that the child's educational needs cannot be met in ordinary neighbourhood schools. However, parents are perfectly entitled to start the ball rolling themselves through discussion with head-teachers, the school medical service, or the local education authority.

The extraordinary thing is that every child I know who has gone to this school has enjoyed great improvement in the eczema within a week or two of starting. The eczema has tended to reappear to a degree during the first half-term, and the first holidays, but thereafter has gradually become more reluctant to recur. The school has an excellent record of academic achievement, and there is an emphasis on sports and other outdoor activities.

The mechanism of the benefit experienced by children who attend such a school is a mystery, though the most likely explanations all centre on the release of the child from an unsatisfactory environment at home. The home environment may be unsatisfactory physically because it contains pets, smokers, or too many house dust mites, or because it is situated in a polluted inner city area. However, I am at least equally concerned about the cumulative problems caused by a home situation that has become too stressful for everyone. When saying this, I am aware that many parents may feel shocked or hurt by such an idea. However, experience suggests that eczema in many of the most severe cases is sustained by tensions within the family situation, both the immediate and the more extended family, and within the community, particularly at school. Liberation can have a surprising effect, both on the eczematous child, and on the rest of the family. The school also helps by providing a climate of confidence. The other children all had severe eczema which has got better, and this helps greatly. Also, these other children help by showing little patience with a fellow pupil who complains unnecessarily about his or her skin, or who attempts to evade any activity because of it. All these factors, together with careful supervision of skin treatment, contribute to the beneficial effect.

While a place at this school will not be appropriate for the vast majority of children with eczema, it should be considered in any child over the age of nine years whose disease is persistently severe despite good medical advice, and whose education has been substantially impaired. For further information, contact the Pilgrims School, Firle Road, Seaford, E. Sussex BN25 2HX (tel. 0323 892697).

CAREERS AND ATOPIC ECZEMA

I am often asked to see student nurses who have developed eczema. It usually turns out that they had eczema when they were children. The eczema later cleared up, as it usually will, but the constant exposure to antiseptics and detergents that is an inevitable part of the job is liable to provoke a recurrence, and sometimes this will mean that nursing has to be given up altogether. It would have been better if this danger had been foreseen and a more suitable career planned at an earlier stage.

The most unsuitable jobs are probably hairdressing and food preparation. These are likely to provoke renewed eczema long after it has initially cleared up, and they will almost invariably aggravate eczema which is already active. Another job which is likely to provoke or aggravate eczema is anyone who has had it in the past is nursing, which I have already mentioned. All dirty jobs may cause problems, for example mechanic, carpenter, bricklayer, painter, or machine tool operator.

Some jobs require a more or less perfect skin. Though some fashion models have been able to become very successful despite having eczema, the stress of the job and worry about the eczema tend to aggravate it, and it is a profession therefore best avoided.

Where the hands are on show, hand eczema may cause real problems, and in many jobs, staff will not be taken on who have hand eczema. A good example is the catering trade, in which hand eczema is seen as hazardous. There is in fact a genuine risk that someone with hand eczema who carries *Staphylococcus aureus* bacteria could be a cause of food poisoning (see p. 72).

The safest jobs are those of a clerical, administrative or professional type, and generally those which require the brain rather than the hands. This is exactly why a good education is really more important for those with eczema than for those without.

13

Eczema and heredity: will our next child get eczema?

Parents of children with severe eczema often ask me if any further children they may have will also be affected. Some parents feel that if the risk were high, they would prefer not to take it. In Chapter 3, we briefly considered the inheritance of the atopic state and, therefore, the predisposition to atopic eczema (p. 10). The situation seems to be a complicated one, and it would be worthwhile looking at it a little more closely.

Let's start by discussing how characteristics are passed on from one generation to the next. Physical differences between people, say the colour of their eyes, result from slight differences in the structure of specific body components, particularly proteins. As we saw in Chapter 5 (p. 44), proteins consist of chains of subunits known as amino acids. There are about twenty different amino acids which can be arranged in an almost infinite variety of sequences, just as the 26 letters of the alphabet can be used to make thousands of different words. The body is able to make up proteins very precisely because it carries a set of exact plans in each and every one of its cells. These plans are called *genes*, and for each and every protein there is a corresponding gene. The genes are arranged in strings known as *chromosomes*. Each gene is similar to written instruction manual, and a chromosome is like a library of such manuals. Each person gets half their genes from one parent and the other half from the other.

Some diseases are inherited in a precise way. A good example is achondroplasia. Achondroplasics are familiar to most people as circus dwarfs, a job for which they are physically particularly well suited. Achondroplasics have a defect in the growth of their bones that causes dwarfing, though in all other ways they are perfectly healthy. The achondroplasia gene contains an error, similar to just one spelling mistake in the instruction manual, which means that an important protein is made slightly wrongly with the result that it doesn't function properly. If an achondroplasic marries a normal person, chances are that half the resulting children will be achondroplasics and the other half will be normal. This is because a child gets half his genes from each parent, so each child has a 50–50 chance of inheriting the achondroplasia gene. This is an example of the simplest pattern of inheritance. Other conditions are passed on in a more complex way. In some, the children will have a one-in-four chance of inheriting the condition; in others, half the boys will inherit it but none of the girls, and so on.

Some characteristics depend on combinations of genes; this is known as 'polygenic' inheritance. In polygenic conditions, the influence of the environment may be very strong, making matters even more complicated. Height is a good example of a characteristic inherited in this way. Height is highly variable, not merely something you have or don't have. A large number of genes determine the potential height of an individual, but environmental factors, particularly nutrition and social conditions, will actually decide whether the potential height is reached. The height of one's parents is important. If one's parents are tall, one will almost certainly be tall; if one's parents are short, one should expect to be short—unless, of course, one's parents are short because of poor nutrition rather than their genes.

The pattern of inheritance of eczema appears to have much in common with the inheritance of height. One inherits the *potential* to develop eczema, not the eczema itself (*Figure 61*). Environmental factors appear to decide whether one actually develops it or not. How bad it is probably also results from a combination of inheritance and environment. If it is true, as we suspect, that between 10 per cent and 20 per cent of all children develop a degree of atopic eczema at some time, then we should perhaps consider it more as a variant of normal than as a disease. It is, of course, only a small proportion of these children in whom it represents a real problem, just as it is in only a small proportion of children that shortness is a real problem.

One's risk of developing eczema can be related, albeit somewhat imprecisely, to the presence or absence of atopic diseases in one's parents—i.e. atopic eczema, asthma, or hayfever. If *one* parent has or has had one of these conditions, the child's risk of developing atopic eczema will be about double that of a child whose parents have never had any of these. If *both* parents have or have had any atopic disease, the risk of eczema in their child is doubled again. In studies done in the last few years, mostly in Scandinavia, where the eczema problem is similar to the UK, atopic eczema was generally reported by parents to have occurred in about 8 per cent of all children by the age of seven years. Where one parent had had any atopic disease, the child's risk of developing eczema was 13 per cent; where both did, the figure reached 18 per cent.

These figures are complicated by the fact that atopic eczema tends overall to be somewhat commoner in boys than in girls.

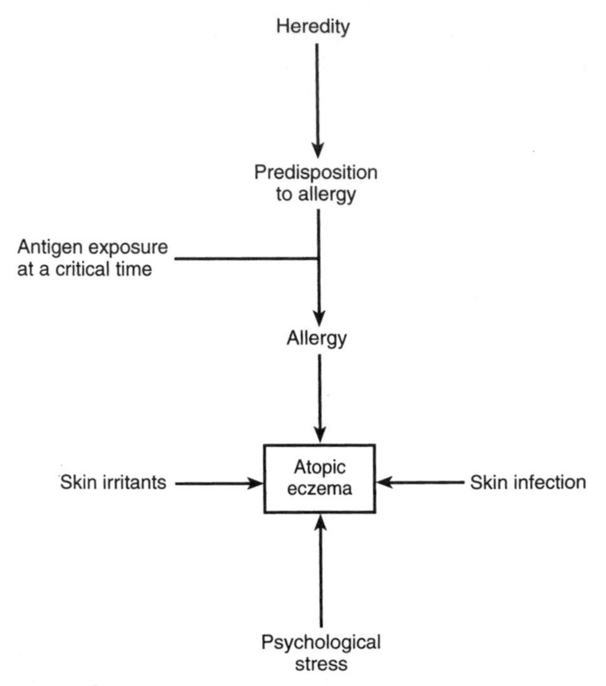

Figure 61 Predisposition to atopic eczema and triggering factors.

Some very recent research has indicated that a child is substantially more likely to develop eczema if the mother is atopic than if the father is. However, though the difference is rather marked if one considers the number of babies developing eczema in the first year of life, it becomes less and less noticeable thereafter.

But what if neither parent has a history of atopic disease, but a brother or sister does? In fact having such a brother or sister increases the risk to about the same degree as having a parent with atopic disease, i.e. about double. Having a brother or sister *as well as* one parent with atopic disease does not appear to increase the risk any further. Figures are not available showing the position if a child has two or more brothers or sisters with atopic disease, but one would imagine that having two or more affected brothers or sisters would be very much like having two affected parents.

If parents, or brothers or sisters, have a history of eczema itself, rather than just asthma or hayfever, then the risk seems to increase

Table 7 The statistical risk of a child developing eczema, and the increased risk if one or two first-degree relatives (brothers, sisters, or parents) themselves have atopic disorders. It can be seen that the risk is highest where relatives have or have had atopic eczema rather than asthma or hayfever.

	Overall for population	No history of atopic disease in family	One parent and/or one brother/sister			Two parents		
			AAD	As/Hf	AE	AAD	As/Hf	AE
Risk of developing atopic eczema (%)	10	8	20	15	25	30	20	40

AAD = all atopic diseases; As/Hf = asthma/hayfever; AE = atopic eczema.

even further (*Table 7*). This implies that in addition to genes transmitting risk of atopic disease in general, there are others which transmit the risk of eczema in particular.

To come back to our original question, what all this means is that if you already have a child with eczema, your next will have about a one-in-four chance of developing it as well. I suspect that many parents worry that the risk may be higher and will find it reassuring that the actual risk is much less than 50:50. It is also important to bear in mind that atopic eczema is usually a fairly mild disease, so that even if their next child is affected, the odds are that the condition will be less severe—another reason to feel assured.

14

Can we do anything to reduce the risk?

Parents who are worried about having a child with eczema often ask if there is anything they can do to reduce the risk. They cannot change the hereditary part of the risk, but it seems logical to anticipate that the environmental contribution could be modified. This possibility is supported by the observation, for example, that risk of eczema differs with month of birth, those children born during the autumn months being at the greatest risk; it has been suggested that this increased risk reflects heavier exposure to house dust mite droppings at this time of year. Whatever the explanation, it seems that events occurring during the first few weeks of life are crucial in determining the likelihood of a genetically at-risk infant actually developing eczema.

FEEDING YOUR NEW BABY

One of the most important factors appears to be the way a baby is fed. There is now a considerable weight of evidence that the risk of developing atopic eczema can be reduced by exclusive breast-feeding for the first few months of life. In the case of babies at high genetic risk, research suggests that feeding nothing other than breast-milk for the first three months of life may reduce the likelihood of eczema to about a half of what it would be if the infant were bottle-fed. Clearly, the effect falls short of complete protection, and many breast-fed babies will go on to get eczema. One consolation if this happens is that breast-fed infants seem to get eczema which is less severe than it would have been, had a different feed been given.

It is not clear whether these benefits of breast-feeding reflect a protective effect of breast-milk, or merely a harmful effect of the usual alternative, cow's-milk-based bottle-feeds. The probability, however, is that it is the foreign proteins in the cow's milk that cause the problem.

Exclusive breast-feeding means no supplementary bottle-feeds at all. Unfortunately, it remains widespread practice in maternity wards to bottle-feed newly born infants at night in order not to disturb the mother. If a baby is delivered by forceps or by caesarean section, it may be taken away from its mother for the first 24 hours, during which time it will usually be bottle-fed. I am glad to say that things continue to change for the better in these respects. Not only is it best for the baby to be breast-fed from the outset, but removal of demand for milk from the breast may actually spoil a mother's chance of ever establishing an adequate supply of milk.

I would urge any mother to plan to breast-feed her baby, but particularly where there is any family history of atopic disease, especially of eczema. Talk to your family doctor, obstetrician, and midwife before the baby is born. Make it clear that breast-feeding is your firm intention, and that you do not wish supplements other than water—which will be given in the form of *dextrose*—to be given to your baby without your permission. If your baby *has* to be given a feed other than from your own breast, where possible express some breast milk in anticipation. *Breast is best* by Dr Penny and Andrew Stanway (Pan Books Ltd, 1983: ISBN 0330281100) provides an excellent introduction to breast-feeding, and I strongly recommend reading it at some time during pregnancy.

What should you do, should breast-feeding be impossible for one reason or another? This is a difficult question to answer. The usual recommendation has been to give a soya formula (p. 164) instead. However, there is in reality very little evidence that soya formula is preferable to cow's milk formula in terms of risk of eczema, and little reason therefore to support its use in preference to cow's milk formula in healthy infants who do not have eczema.

If soya formula is no better than cow's milk formula, what about hydrolysate formula, such as *Nutramigen® or *Pepti-Junior®? As we considered earlier (p. 164), these are usually made from protein obtained from cow's milk. However, during the manufacturing process, this protein is *hydrolysed*, which means that it is broken down to very short chains of amino acids. This process substantially reduces the protein's ability to cause allergy and the feed is, theoretically, more or less non-antigenic. Nutritionally, these hydrolysate formula feeds are made to resemble human milk as much as possible. They should therefore be an ideal alternative to breast milk where one is needed, but their high cost is a major snag. For this reason, they should only be used when the risk is a high one, and on the recommendation of your family doctor, obstetrician, or paediatrician.

If possible, you should aim at feeding your baby nothing other than your own milk for the first four months of its life. The protective effect may be even greater if you continue breast-feeding for longer than this, though you should be aware that human milk alone is not a nutritionally adequate diet for babies older than about six months.

There are probably many reasons why exclusive breast-feeding is not a complete protection against atopic eczema. One is the difficulty of ensuring that breast-feeding is genuinely exclusive. I have already drawn attention to the still all-too-common practice in hospitals of giving babies extra bottle-feeds during their first hours of life. This practice may be well-meant, but it is ill-advised, and could nullify the benefits of subsequent exclusive breast-feeding.

MOTHER'S OWN DIET

Another reason is that food antigens from a mother's own diet are transmitted, albeit in minute amounts, to her baby in her own milk. It may seem incredible, but there is no doubt that this does occur. After a lactating mother drinks cow's milk, for example, intact cow's milk proteins can be found in her milk. That this could provoke allergy is demonstrated by the fact that many exclusively breast-fed babies develop antibodies to these cow's milk proteins before a drop of cow's milk has crossed their lips. Although such antigens are present in breast milk in quantities sufficient to induce a response in the baby's immunological system, the quantities are small and the full consequences may therefore not be seen until the child starts to eat the relevant foods after weaning. For this reason, the consequence of this *sensitization* during breast-feeding may be the appearance of eczema during weaning. Indeed, it is quite common for parents to report that their baby first developed eczema, or some other manifestation of cow's milk allergy, within one or two days of having their first cow's milk feeds. This is much too quick for the allergy to have been induced by these first feeds, and it implies that such babies were already sensitized.

It is therefore a logical question whether breast-feeding mothers should eliminate cow's milk and perhaps certain other foods from their own diet. Although the principle is a sound one, the best way to put it into practice is not established since no-one knows for certain which foods should be avoided, when this avoidance should start, or how strict it must be. I can only give what I would regard as a sensible approach to this issue.

First of all, I do not think that any mother-to-be should contemplate eliminating foods from her diet while breast-feeding unless her child is at high risk of developing eczema. The higher the risk, the more worthwhile such a manoeuvre is likely to be. For the present, it is only in relation to eggs and cow's milk that a strong case can be made out for avoidance.

If eggs and cow's milk are avoided by breast-feeding mothers, adequately nutritious alternatives *must* be incorporated into the diet in compensation. This means alternative sources both of first-class protein and calcium. As we considered in an earlier chapter (p. 64), most shop-bought soya 'milks' contain little protein and virtually no calcium. The easiest way to replace the protein content of milk and eggs is to eat more meat and/or fish. If you are a vegetarian and you don't eat fish either, you may have problems obtaining an adequate amount of first-class protein for breast-feeding if you are also not eating eggs or milk. It is possible, however, but only by knowing a great deal about this type of diet (technically a *vegan* diet), and by eating a good quantity of pulses and nuts. In view of the frequency of peanut allergy in eczematous children, this may however not be a very good idea.

If the calcium content of the mother's normal milk intake is not replaced, it is very unlikely that the diet will contain enough from other sources. The result will be that calcium will be extracted from the mother's bones and teeth to maintain the supply in her milk for as long as possible, a process which poses a major threat to her long-term health in terms of increased bone fragility. Some soya 'milks' have adequate amounts of calcium added to them, but failing this, the most reliable way to ensure sufficient calcium intake is to take two effervescent *Sandocal® tablets daily. Ideally, this regimen should be started a few days before delivery, and continued throughout lactation. In order to succeed in avoiding egg and cow's milk protein completely, I would very strongly recommend that advice be sought from a dietitian; this can be arranged by your GP. However, I doubt whether it is necessary for the avoidance to be obsessional for it to be beneficial, and I imagine that the very small amounts of these proteins—found, for example, in biscuits and cake—would be more or less irrelevant.

The question obviously arises whether similar precautions should also be taken during pregnancy. There is evidence that some babies are already sensitized to foods, particularly eggs and cow's milk, at birth, which implies that food antigens can cross the placenta from the mother's bloodstream. An attempt to reduce one's intake of these items, perhaps during the last six months of pregnancy, might therefore seem as logical as doing so during lactation. Research has been conducted to look at this and, for the moment, the evidence suggests that milk and egg elimination during pregnancy *does not* reduce the subsequent risk of eczema, whereas the same manoeuvre during breast-feeding definitely does seem to do so. On the other hand, it is not sensible or necessary to increase your intake of either food during pregnancy, and if your diet was previously adequate, you should resist pressure to do so.

WEANING

Introduction of some solids should ideally be instituted no earlier than the fourth month, but no later than the seventh. The beginning of the fifth month is probably the ideal time, if you can hold on.

Our present state of knowledge does not allow very precise advice to be given, but on balance it seems wise to avoid giving an at-risk baby any cow's milk until the age of at least one year, and egg or nuts until at least two years. Cow's milk avoidance implies avoidance of all foods containing cow's milk protein, which means keeping a careful eye on the contents of prepared baby foods. Any foods containing whey or casein should therefore not be used.

You should aim to breast-feed after each solid feed until the end of the first year. If you cannot do this, a hydrolysate formula may be the best alternative.

The first nutritional components your baby will need in addition to your own milk are fibre, to improve bowel function, and carbohydrate. Your milk, or hydrolysate formula, will continue to provide more than enough protein and fat at least until the end of the first year.

For fibre, you need to introduce vegetables and fruits first of all. Suitable vegetables include carrots, swedes, turnips, peas, beans, cabbage, broccoli, Brussels sprouts, and cauliflower. These should be boiled carefully (because overcooking destroys much of their nutritional value) and then mashed. A little milk-free margarine—for example, Tomor®—may be added for its softening effect. Suitable fruits include mashed banana, stewed apple or pear, or mashed tinned fruits such as peaches or apricots.

Potato should be preferred to cereal at this stage as a source of carbohydrate. It is a good idea to mix boiled mashed potato with other vegetables. Give such feeds three times daily, finishing off with a feed of human milk or hydrolysate formula.

After a month, you could introduce a suitable cereal. I think it might be a good idea to avoid wheat-based cereals during the first year, but I should point out that there is little evidence in favour of such a view. Nevertheless, I prefer to recommend baby rice, or, perhaps better still, sago or tapioca (*not* semolina, which is made of flour) at this stage.

This combination of foods is enough until your child is a year old, when you should introduce meat. Turkey, pork, and lamb are good choices as there is no overlap of proteins with egg and cow's milk. Serve with vegetables and potato. At this stage, you could also try fish, which should initially be white fish such as haddock or cod.

In the second year, you could attempt gradual introduction of cow's milk, which can be in the form of ordinary doorstep milk. Full fat (silver-top) milk is recommended for children up to the age of five years, though semi-skimmed (red-top) is satisfactory after the age of two years where this is in general use in a household. From the age of five years, semi-skimmed milk is probably ideal, and is preferable to fully skimmed milk. Rub some milk on to your child's lip and on to a cheek before giving a very small amount by mouth if no reaction occurs (see p. 39). Continue to avoid eggs and nuts.

NON-FOOD ANTIGENS

Babies are of course exposed to antigens other than just those in their food. They come into contact with airborne antigens the same as everyone else, and these include, as we have discussed earlier, pollen spores, animal dander, and house dust mite droppings. As I mentioned, there is evidence that babies born in the autumn, when house dust mites are most populous, may be more likely to develop eczema than babies born at any other season. For this reason, it seems wise to make efforts to reduce the population of house dust mites in the immediate environments of an at-risk baby (see p. 174). There is also evidence that early exposure to furry pets increases the risk of subsequent allergy, and it is undoubtedly wisest not to have a dog or cat in a home where there is an at-risk baby.

OTHER PRECAUTIONS

The most important additional protection for your baby is to avoid exposing it to tobacco smoke.

CONCLUSIONS

There is for the present no way of guaranteeing prevention of eczema by the kind of manoeuvres we have been discussing. Nevertheless, in the light of our current knowledge, these precautions seem sensible, and no doubt they can lead to some reduction of risk. Perhaps our greatest hopes for the future lie in the area of prevention, as our ability to cure atopic eczema, once it is established, is likely to remain rather limited.

Appendices

APPENDIX 1: THE NATIONAL ECZEMA SOCIETY

If your child has eczema or if you have eczema yourself, much can be gained through contact with other parents and sufferers. A few years ago, some patients, and doctors got together and formed a society, which they called the National Eczema Society, whose object was to help sufferers, and those who care for them, to help themselves. The society attempts to achieve this in several ways. First, it provides opportunities for people with eczematous children or partners, or with eczema themselves, to meet. Great comfort can come through sharing experiences and the feeling of no longer being alone. A society of this type can also help by making the problem of eczema better known to people generally, perhaps making it less mysterious, less threatening, and less misunderstood. A more sympathetic attitude in the general public would help enormously. Not least among the aims of the society are those of improving the treatment and understanding of eczema; these are accomplished by interesting both doctors and scientists in the problem of eczema, and by financing specific research projects.

The National Eczema Society also publishes a magazine called *Exchange* four times a year; this carries articles by doctors and other health professionals and also provides an exchange of information between members. Regular meetings provide the opportunity to hear professionals talk in a down-to-earth way (one hopes!) about eczema and the problems it causes, and interesting discussion often follows. The society has an office which is staffed during working hours, and from which information of all sorts can be obtained, at:

4 Tavistock Place
London WC1H 9RA
Tel. 071 388 4097

APPENDIX 2: SOME RECOMMENDED BOOKS

Breast is best, by Dr Penny and Andrew Stanway, Pan Books Ltd, 1983, £4.99: ISBN 0330281100

New 'E' for additives, by Maurice Hanssen, Thorsons, 1987, £4.50: ISBN 0722515626

Look at the label, MAFF Publications Unit, Food Standards Division, Ergo House, Smith Square, London SW1, tel. 071 270 3000 or 071 238 6260. Single copies free.
National Eczema Society Information Packs (see Appendix 1).

APPENDIX 3: SOME USEFUL ADDRESSES

Air Improvement Centre (humidifiers)
23 Denbigh Street
London SW1V 2HF
Tel. 071 834 2834

Allerayde (vacuum filters, mattress and pillow covers, tannic acid spray)
147 Victoria Centre
Nottingham NG1 3QF
Tel. 0602 240983

Alprotec (Allergen barrier bedcovers)
Advanced Allergy Technologies Ltd,
Royd House
224 Hale Road,
Altrincham, Cheshire WA15 8EB
Tel. 061 903 9293

Angela Knitwear (cotton pullovers)
Ferndale
Manor Road
Writhlington
Radstock
Bath BA3 3LZ

Bel Air Associates (air filtration units)
63 Oakway
Woking
Surrey GU21 1TS
Tel. 04867 3931

Bio-Diagnostics Ltd (vinyl gloves, disposable vinyl gloves, cotton gloves)
Upton Industrial Estate
Rectory Road
Upton upon Severn
Worcs WR8 0XL
Tel. 06846 2262

Cotton-On Ltd (cotton clothing, mitts)
Monmouth Place
Bath BA1 2NP
Tel. 0225 461155

100 Per Cent Cotton (made-to-measure clothing)
22 Hambledon Court
Holmwood Gardens
Wallington
Surrey SM6 0HN
Tel. 081 669 6028

The Family Fund
PO Box 50
York YO1 2ZX

Foodwatch (special dietary products—mail order)
Butts Pond Industrial Estate
Sturminster Newton
Dorset DT10 1AZ
Tel. 0258 3356

W. L. Gore and Associates Ltd (Intervent Bedding System)
Church Gate
Church Street West
Woking
Surrey GU21 1DJ
Tel. freephone 0800 515730

Kids' Stuff (clothing)
10 Hensmans Hill
Bristol BS8 4PE
Tel. 0272 734980

Medic-Alert Foundation (bracelets and pendants warning of allergies or steroid treatment)
12 Bridge Wharf
London N1 9UU
Tel. 071 833 3034

Medivac Healthcare Products (bedding products, vacuum cleaners)
Bollin House
Riverside Works
Manchester Road
Wilmslow
Cheshire SK9 1BJ
Tel. 0625 539401

National Eczema Society
4 Tavistock Place
London WC1H 9RA
Tel. 071 388 4097

National Safety Associates of America (UK) Ltd (air filtration systems)
NSA House
1 Reform Road
Maidenhead
Berks SL6 8BY

Nutricia Dietary Products (special dietary products—mail order)
494–96 Honeypot Lane
Stanmore
Middx HA7 1JH
Tel. 081 951 5155

Permutit water softeners, Ecowater Systems Ltd
Unit 1, The Independent Business Park
Mill Road
Stokenchurch
Bucks HP14 3TP
Tel. freephone 0800 521143

Pilgrims School
Firle Road
Seaford
E. Sussex BN25 2HX
Tel. 0323 892697

Porvair Ltd (vacuum filters)
Estuary Road
Riverside Industrial Estate
King's Lynn
Norfolk PE30 2HS
Tel. 0553 761111

Slumberland p.l.c. Medicare Division (Health Seal range of bedding products)
Salmon Fields
Oldham OL2 6SB
Tel. 061 628 5293

Taylormaid Products Ltd (Medivac vacuum cleaners, Banamite)
18A Water Lane
Wilmslow
Cheshire SK9 5AA
Tel. 0625 539401

Vorwerk UK Ltd (Vacuum cleaners)
Toutley Road
Wokingham
Berks RG11 5QN
Tel. 0734 794753

Index

Proprietary (trade) names are printed with initial capitals